Past Masters
General Editor Kei

Lamarck

Past Masters

AQUINAS Anthony Kenny
ARISTOTLE Jonathan Barnes
BACH Denis Arnold
FRANCIS BACON Anthony Quinton
BAYLE Elisabeth Labrousse
BERKELEY J. O. Urmson
THE BUDDHA Michael Carrithers
BURKE C. B. Macpherson
CARLYLE A. L. Le Quesne
CHAUCER George Kane
CLAUSEWITZ Michael Howard
COBBETT Raymond Williams
COLERIDGE Richard Holmes
CONFUCIUS Raymond Dawson
DANTE George Holmes
DARWIN Jonathan Howard
DIDEROT Peter France
GEORGE ELIOT Rosemary Ashton
ENGELS Terrell Carver
GALILEO Stillman Drake
GOETHE T. J. Reed
HEGEL Peter Singer
HOMER Jasper Griffin
HUME A. J. Ayer
JESUS Humphrey Carpenter
KANT Roger Scruton
LAMARCK L. J. Jordanova
LEIBNIZ G. MacDonald Ross
LOCKE John Dunn
MACHIAVELLI Quentin Skinner
MARX Peter Singer
MENDEL Vitezslav Orel
MONTAIGNE Peter Burke
THOMAS MORE Anthony Kenny
WILLIAM MORRIS Peter Stansky
MUHAMMAD Michael Cook
NEWMAN Owen Chadwick
PASCAL Alban Krailsheimer
PETRARCH Nicholas Mann
PLATO R. M. Hare
PROUST Derwent May
TOLSTOY Henry Gifford

Forthcoming

AUGUSTINE Henry Chadwick
BERGSON Leszek Kolakowski
JOSEPH BUTLER R. G. Frey
CERVANTES P. E. Russell
COPERNICUS Owen Gingerich
DESCARTES Tom Sorell
DISRAELI John Vincent
ERASMUS John McConica
GIBBON J. W. Burrow
GODWIN Alan Ryan
HERZEN Aileen Kelly
JEFFERSON Jack P. Greene
JOHNSON Pat Rogers
KIERKEGAARD Patrick Gardiner
LEONARDO E. H. Gombrich
LINNAEUS W. T. Stearn
MILL William Thomas
MONTESQUIEU Judith Shklar
NEWTON P. M. Rattansi
ST PAUL G. B. Caird
ROUSSEAU Robert Wokler
RUSKIN George P. Landow
RUSSELL John G. Slater
SHAKESPEARE Germaine Greer
ADAM SMITH D. D. Raphael
SOCRATES Bernard Williams
SPINOZA Roger Scruton
VICO Peter Burke
VIRGIL Jasper Griffin
WYCLIF Anthony Kenny
and others

L. J. Jordanova

Lamarck

Oxford New York
OXFORD UNIVERSITY PRESS
1984

Oxford University Press, Walton Street, Oxford OX2 6DP
London New York Toronto
Delhi Bombay Calcutta Madras Karachi
Kuala Lumpur Singapore Hong Kong Tokyo
Nairobi Dar es Salaam Cape Town
Melbourne Auckland

and associated companies in
Beirut Berlin Ibadan Mexico City Nicosia

Oxford is a trade mark of Oxford University Press

© L. J. Jordanova 1984

First published 1984 as an Oxford University Press paperback
and simultaneously in a hardback edition

All rights reserved. No part of this publication may be reproduced,
stored in a retrieval system, or transmitted, in any form or by any means,
electronic, mechanical, photocopying, recording, or otherwise, without
the prior permission of Oxford University Press

This book is sold subject to the condition that it shall not, by way
of trade or otherwise, be lent, re-sold, hired out or otherwise circulated
without the publisher's prior consent in any form of binding or cover
other than that in which it is published and without a similar condition
including this condition being imposed on the subsequent purchaser

British Library Cataloguing in Publication Data

Jordanova, L. J.
Lamarck.—(Past masters)
1. Lamarck, J. B. 2. Biology—History
I. Title II. Series
574'.092'4   QH31.L2
ISBN 0-19-287588-4
ISBN 0-19-287587-6 Pbk

Library of Congress Cataloging in Publication Data

Jordanova, L. J.
Lamarck.
(Past masters)
Bibliography: p.
Includes index.
1. Lamarck, Jean Baptiste Pierre Antoine de Monet de 1744–1829.
2. Naturalists—French—Biography
I. Title. II. Series.
QH 31.L2J67 1984   574'.092'4 [B]   84-5697
ISBN 0-19-287588-4
ISBN 0-19-287587-6 (pbk.)

Set by Hope Services, Abingdon
Printed in Great Britain by
Cox & Wyman Ltd, Reading

For my mother
Phyllis Jordanova

Preface

Reactions to Lamarck have always been mixed. Today the most common response to his ideas is to dismiss them as simply wrong. Historians, however, have a duty not to be drawn into making such absolute judgements and instead to provide accounts of thinkers like Lamarck which make their ideas accessible and comprehensible. My purpose has therefore been to set out Lamarck's theories of nature in general and of the living world in particular, in their own terms. There are two reasons why this is more helpful than a catalogue of Lamarck's 'right' and 'wrong' beliefs. The first is that Lamarck was a thinker in the traditions of the late Enlightenment and his ideas about the history of nature must be understood in that context. The second is that Lamarck's name was frequently invoked in the nineteenth- and early twentieth-century debates about evolutionary theories. To dismiss his notions as 'wrong' would be to neglect the power they indisputably had over later generations.

During his long life Lamarck became involved with subjects which may now strike us as surprising when we consider that his main contributions were to biology. Yet it would be wrong to neglect his chemical speculations or his meteorological observations. I argue that these areas shed important light on his thought and contribute to our understanding of it as an integrated whole. The main concern which united Lamarck's apparently disparate interests was scientific method. He had a lively appreciation of the difficulty of unravelling nature's processes. The account which follows treats Lamarck's work as a subject in the

Lamarck

history of ideas and attempts to reconstruct his intellectual adventures in searching for a conceptual basis for biology, the science of living things.

In the course of my work on Lamarck I have incurred many debts and I would like to thank all those who have assisted me. Earlier versions of this book have been read, in whole or in part, by Joyce Appleby, Karl Figlio, Angela Livingstone, Roger Smith and Charles Webster. Their comments and suggestions have been of immense value; I am grateful to them not only for generously taking time to help, but also for their enthusiastic encouragement. I wish to acknowledge the kindness of Pietro Corsi and Dorinda Outram in allowing me to see unpublished work; of Chip Burkhardt in sharing rare Lamarck materials with me; of Don Smith in advising on matters botanical; of Roger Huss in helping to translate Lamarck's French into comprehensible English and of the Royal Society in providing financial support for research on Lamarckism.

I was a guest of the Wellcome Unit for the History of Medicine at the University of Oxford while writing the first draft of the book. It is a pleasure to record my gratitude to its members for the intellectual stimulation and friendly comradeship they offered. Without the prompt and helpful services provided by the Inter-Library Loan Department at the University of Essex, my task of preparing the final draft would have been considerably harder.

At Oxford University Press I had an exceptionally patient and helpful editor from whom I have learnt much and to whom I am greatly indebted. On a more personal note: special thanks to Joan Busfield and Steve Smith for their friendship and support; and, above all, to Karl Figlio and to our daughter Sonya, who make everything possible. I owe so much to my mother that I dedicate this book, with thanks, to her.

L. J. JORDANOVA

Contents

Abbreviations x
1 Lamarck's life 1
2 Classification and scientific method 11
3 Plants 25
4 Animals 33
5 Life 44
6 The science of the environment 58
7 Transformism and the *Zoological Philosophy* 71
8 Nature and God 83
9 Man 89
10 Lamarck's legacy 100

Further reading 114

Index 116

Abbreviations

A   *Histoire naturelle des animaux sans vertèbres*, 7 vols, Paris, 1815–22 (*Natural History of Invertebrates*)

D   *Nouveau Dictionnaire d'histoire naturelle*, Paris, 1817–18 (*New Dictionary of Natural History*)

E   *Encyclopédie méthodique-botanique*, 4 vols, Paris, 1783–95 (*Botanical Dictionary*)

H   *Hydrogéologie*, Paris, 1802 (*Hydrogeology*, edited by A. Carrozzi, Urbana, Illinois, 1964; all quotations from this edition)

M   *Annuaires météorologiques*, 11 vols, Paris, 1799–1810

N   'Lamarck in 1800. A Lecture on the Invertebrate Animals and a Note on Fossils taken from the *Système des animaux sans vertèbres*', translated and annotated by D. R. Newth, *Annals of Science*, 1952, vol. 8, 229–54

P   *Philosophie zoologique*, 2 vols, Paris, 1809 (*Zoological Philosophy*)

R   *Recherches sur l'organisation des corps vivans*, Paris, 1802 (*Researches on the Organisation of Living Bodies*)

S   *Système analytique des connoissances positives de l'homme*, Paris, 1820 (*Analytical System of Man's Positive Knowledge*)

Quotations in the text are in my own translation, except for those denoted by H or N.

1 Lamarck's life

Lamarck put forward the first systematic account of the historical development of living nature. He showed that the past is continuous with the present, making the present both the key and heir to the past. In demonstrating that natural forms had changed over time, Lamarck drew on a number of scientific disciplines—botany, zoology, geology, meteorology, chemistry, psychology. The term 'biology' was coined by Lamarck to denote a separate science devoted to living things, and he continually emphasised the distinctiveness of the organic world.

Organisms could best be understood, he found, in terms of their interaction with and adaptation to the environment. It followed that biology could not be studied in isolation from those sciences which explained the physical world. In defining the new discipline of biology, Lamarck anticipated that those features common to plants and animals would be investigated. He confidently defined man, in both his physical and mental aspects, as part of nature, thereby repudiating any religious and philosophical claims for free will and an immortal soul. He contributed to the extension of the domain within which the natural sciences could legitimately operate. The possibilities for the biological and human sciences which were thereby opened up were only fully grasped by later generations.

The legacy which Lamarck bequeathed to the naturalists who followed him was in part a methodological one. From his early work on taxonomy, the science of classification, he concerned himself with the means by which the natural world could best be understood. To this end he continually elaborated models and analogies of processes which could

Lamarck

not be observed directly. The natural sciences have always depended on such devices, and by his persistent attempts to visualise nature's processes, Lamarck made it possible for others to think more critically and productively about what the natural world was like. His achievement was, on the one hand, to define an expanded field of scientific operation and, on the other, to attempt an imaginative reconstruction of nature's ways.

Lamarck's name is invariably associated with the concept of the inheritance of acquired characteristics (usually called 'characters' by biologists). The idea was not original to him; rather it had been for centuries, and continued to be into the twentieth century, a commonplace. While it certainly played a role in Lamarck's theories of biological change as one of his postulated mechanisms of organic variation, its importance has been much exaggerated. This happened because, as chapter 10 explains, the inheritance of acquired characters became the subject of vociferous debate at the end of the nineteenth century. Lamarck's beliefs are best understood in the context of their own times instead of being viewed through the spectacles of later commentators.

Ideas associated with Lamarck's name continued to arouse passionate reactions from biologists well into the twentieth century. However speculative his ideas may appear to be in retrospect, they stimulated the imaginative drive for an integrated worldview. They appealed especially to those who opposed the analysis of organisms in terms of physics and chemistry (reductionism) and to those who sought meaningful patterns and purpose in nature. His legacy was in important respects a literary one, deriving from his search for an adequate scientific language through which biological processes could be analysed. He touched on many issues to which there are no easy solutions: the relationship between mind and body, the role of behaviour in evolution, and the extent to which organisms can adapt

Lamarck's life

to their environment. Lamarck saw the potential for an integrated science of living things when many others, under the sway of Cartesian mechanism, held that organisms were merely complex machines to be understood, like all nature's puzzles, through the physico-chemical sciences. Like his contemporary the great anatomist Xavier Bichat (1772–1802), Lamarck was sensitive to the delicate conceptual problems posed by biology. To marry this recognition of the distinctive scientific challenge which biology offered with an understanding that nature was a developing, historical phenomenon was a significant intellectual achievement. If biologists have returned repeatedly to Lamarck's ideas, it is because they are seen to be capable of nurturing a variety of scientific approaches to the mysterious phenomenon of life.

Jean-Baptiste Pierre Antoine de Monet de Lamarck was born into the minor nobility in Bazentin, Picardy, on 1 August 1744. He was educated at the Jesuit College of Amiens between about 1755 and 1759, forming an early ambition to become a soldier, which he did during the Seven Years War by leaving home on an old nag and joining up. Anecdotal evidence suggests that he distinguished himself as a soldier by his bravery and his tenacity. Tenacity was at any rate a quality he displayed throughout his life. When in 1768 an injury forced him to abandon a military career, he settled in Paris, arriving there probably in 1769 or 1770. During his early years in the capital, he worked in a bank and attended lectures, including medical ones.

As a young adult Lamarck had two enthusiasms, botany and meteorology, which were to become of great significance in his life. Out of his pleasure in collecting plants came his first masterpiece, the definitive *French Flora* (*Flore françoise*), published in 1779. His passion for

Lamarck

observing the skies developed into a particular concern for meteorology as a new scientific discipline. Although Lamarck's writings on the subject—especially his meteorological annuals, published between 1799 and 1810—were poorly received by contemporaries, this science gave him valuable insight into the impact of atmospheric and, more generally, environmental changes on living things. That insight informed his whole conception of biology and biological change.

Lamarck was lucky enough to receive patronage from the great naturalist Georges, comte de Buffon (1707–88), director of the Jardin du Roi in Paris. This garden housed natural history collections, and became, as it still is today, the major centre for the study of natural history in France. Lamarck spent most of his working life there and benefited from its unique collections, which vastly increased during his career, largely thanks to plundering by Napoleon's armies. It was Buffon who helped Lamarck to have the *French Flora* published by the government. The fashionable status botany then enjoyed ensured the work a favourable reception. It was both accurate and simple to use. Lamarck had correctly anticipated that the new *Flora* would prove a valuable and popular enterprise.

In 1779 Buffon engineered Lamarck's election to the Académie Royale des Sciences, a distinguished body which gave its members a State pension. Lamarck shared with many other academicians a belief in the public utility of the natural sciences—a belief portrayed in the Academy's reports on matters of interest and use to the State. He faithfully attended meetings until old age was finally too great an obstacle. Although a member of the botany section, he presented many papers on chemistry, physics and meteorology. Generally they were coolly received, and this lack of enthusiasm made him bitter, which in turn led to strained relationships between him and some of his peers.

Lamarck's life

Yet, Lamarck was by no means isolated from the intellectual life of Paris. He was friendly, for example, with others who worked on invertebrate animals, like Olivier (1756–1814), Bruguières (1750–99) and Latreille (1762–1833). Among older savants, the crystallographer Haüy (1743–1822) offered him help and encouragement. Étienne Geoffroy Saint-Hilaire (1772–1844) certainly felt a kinship with Lamarck, and even Blainvaille (1777–1850), at one time a protégé of Cuvier (1766–1832), with whom Lamarck was not on friendly terms, attempted to give his achievements due recognition. Lamarck's merits were thus apparent even to those who did not share his views. Later, when the general scientific atmosphere, especially in France, was more in sympathy with Lamarck's ideas, naturalists like Alfred Giard (1846–1908) and Edmond Perrier (1844–1921) strove to make his work understood and respected.

Lamarck lost sympathy because he refused to abandon views that were resolutely opposed to innovations which others saw as dramatic and impressive advances. In some cases his opposition has been vindicated; in others it has not. Whatever the ultimate outcome of scientific disputes in his time, Lamarck's willingness to stand alone in defence of his beliefs was an outstanding characteristic of his life and career. When his enemies achieved positions of power during and after the Revolution of 1789, they did not refrain from voicing publicly their lack of respect for him and his ideas. That others continued to support him is an indication of the variety of scientific theories and methods then current.

In 1781 Lamarck acted as tutor and companion to Buffon's son Georges during his European travels. Georges was a difficult young man by all accounts, and the job was evidently not to Lamarck's taste; but it did give him what proved to be his only chance to travel and to examine plants and minerals in areas unfamiliar to him. Between 1788 and

Lamarck

1793, Lamarck held a number of minor positions involving botanical work at the Jardin du Roi. In 1793 he and his colleagues proposed a plan for the reform of their institution into the Muséum d'Histoire Naturelle, by the foundation of twelve equal professorships to cover the various aspects of the natural world. Lamarck was assigned to study the insects and worms, a miscellany of simple animals in which others had little interest. We do not know whether he asked for the post, or whether it was given to him because no one else cared to take it on. The attention he consequently gave to the animals he was to name invertebrates had a profound effect on his approach to natural history.

The most dramatic changes in Lamarck's career were in the 1790s. He published three treatises devoted to physics and chemistry which contained his germinating ideas about a science of life and an examination of such phenomena as colour and heat. His theories about the nature of living things and about the physical world were intimately linked. Also developing fast in this period were his zoological researches on fossil shells, for which Lamarck drew on his private collection as well as those of the Muséum. His writings on botany and palaeontology won praise and approval from the scientific community, unlike his work in chemistry, physics and meteorology.

Lamarck's work with fossils led him to consider the possible changes organisms had undergone during the history of the earth. From there he went on to examine the geological formations in which fossils were found and he became increasingly convinced of the importance of looking at the environment to elucidate its impact on organisms. His first published definition of 'biology', the new scientific discipline he proposed for the study of the living world, appeared, appropriately, in his *Hydrogeology* (*Hydrogéologie*, 1802); for it was by stressing that all changes on the earth

Lamarck's life

took place slowly, principally through the agency of water, that Lamarck set the stage for a view of nature, both living and inert, as the product of an immensely long past. This proposition lay at the core of his evolutionary theories.

The revolutionary period was an important one in Lamarck's life. There is no evidence to suggest that he suffered any personal hardship at this time. As a child of the French Enlightenment, he was sympathetic to the aims of the Revolution, believing in human progress, the dignity of mankind, and the power of reason to overcome social and political oppression. His membership of the liberal Société de 1789 suggests that he embraced ideas of citizenship, equality, liberty and brotherhood and supported moderate, rationally conceived plans for reform. Yet in his later years Lamarck was deeply pessimistic about the future of the human race and its planet, repeatedly voicing bitterness and despair in his writings.

The professors at the Muséum were required to give courses of lectures for students on their assigned subjects. Lamarck offered an annual course on invertebrate zoology from 1794 until 1820; thereafter Latreille, the noted entomologist, deputised for him. He lectured to a mixed audience comprising students of many nationalities and of a variety of ages. Many had medical interests. Apart from the first lecture, the course was arranged on strict taxonomic lines. We can trace the development of Lamarck's ideas through his annual introductory lectures. The lecture in 1800 contained the first public statement of his new-found conviction that species, like everything else in nature, are subject to change. By the closing years of the eighteenth century, Lamarck was increasingly convinced that species had not been created all at once and for ever, as most people believed, but that they had evolved gradually over time by entirely natural processes which it was the business of science to reveal.

Lamarck

We may properly call the theory Lamarck subsequently proposed 'transformist', since this is the literal translation of the French term *transformiste*, used in nineteenth-century France to denote both Darwinian and Lamarckian theories. (In French *évolution* had many connotations but none of them corresponded to the English usage.) 'Transformism' serves to remind us of the distinctiveness of Lamarck's ideas and of general approaches to biology current in nineteenth-century France. Where evolution suggests natural selection which emphasised struggle for survival and competition, transformism evokes the continuous adaptive changes and harmonious balance of nature which Lamarck envisaged.

The theoretical contribution made by Lamarck's transformism was to explain a wide range of biological phenomena in a coherent and economical fashion. His main ideas were sketched out in 1800 in *A System of Invertebrate Animals* (*Système des animaux sans vertèbres*), and more fully developed two years later in *Researches on the Organisation of Living Bodies* (*Recherches sur l'organisation des corps vivans*). But his *magnum opus* was *Zoological Philosophy* (*Philosophie zoologique*, 1809), an impressive synthesis embracing a new classification of the animal kingdom, transformism, a natural history of invertebrates and a detailed account of the operations of the nervous system in a wide range of animal groups. This was and remains his most widely read book. When it was published Lamarck was already sixty-five.

Lamarck continued to write until he was well into his seventies. Between 1815 and 1822 appeared his *Natural History of Invertebrates* (*Histoire naturelle des animaux sans vertèbres*), a work which had been eagerly awaited, even by those who remained unimpressed by his philosophical system. It demonstrated once again his skill as a practical naturalist as well as his synthetic vision of the

Lamarck's life

natural sciences. It would be a mistake to separate Lamarck's success in taxonomy and descriptive natural history from his theoretical inclinations. For him they went together, as is clearly shown in *An Analytical System of Man's Positive Knowledge* (*Système analytique des connoissances positives de l'homme*, 1820). It was concerned with metaphysics and the basis of human knowledge, and it reveals the distinctive character of his thought. It is his least self-conscious book, a work by a man of seventy-six permeated with the spirit he brought to science—a sense that nature was a unified system of laws, laws to be inferred from the observation of natural phenomena.

Although we know very little about Lamarck's private life and his personality, there can be no doubt that he was not a particularly conventional man. He lived for many years with a woman who bore him six children, marrying her only on her death-bed. He married again twice, possibly three times and had two more children by his second wife. He was a man of great passion who fought tenaciously for dearly held beliefs regardless of fad or fashion. Lamarck was no literary stylist. His lengthy, convoluted sentences and extensive repetitions are surprising in view of his lifelong preoccupation with accurate nomenclature, precise definition and clear, unambiguous language.

It is consistent with the little we know of Lamarck as a person that he gave his readers few clues as to his sources and intellectual debts. He was clearly familiar with the writings of the *philosophes* and he had a high regard for Rousseau. Naturally he was also well versed in scientific, medical and philosophical works by such people as Buffon, Linnaeus, von Haller, Locke, Descartes (1596–1650) and Bonnet (1720–93). The catalogue of Lamarck's library indicates that he kept up to date with new work. There is no evidence, however, to show that he knew the ideas of Erasmus Darwin (1731–1802), similar though they were in

Lamarck

many ways to his own. In his published work ideas are presented as if they have just sprung, fully developed, from Lamarck's own head.

This reticence has heightened the feeling that Lamarck was an isolated visionary, almost an outcast. There is little historical evidence to support such a view. He was active in a number of learned societies. His papers and books were regularly discussed in the *Moniteur Universel* and the *Magasin Encyclopédique*, respectively the State newspaper and a prominent literary journal. He participated in two important collective publishing ventures; the continuation of the great *Encylcopedia* of Diderot and D'Alembert, for which he wrote a substantial part of the botany section (1783–95), and a natural history dictionary in which his articles appeared in 1817 and 1818. In the last twenty years of his life, Lamarck progressively lost his sight; by 1818 he was totally blind and dependent on his daughters. He died in poverty on 18 December 1829.

Lamarck was an active scientist for more than forty years, during which time his ideas about the living world underwent significant changes. The most important of these was his gradual conversion during the 1790s to a transformist position. We know relatively little about the historical circumstances which led to this shift. Indeed far less is known about Lamarck's life than about the lives of most other scientists of comparable stature. Those who write about Lamarck must therefore learn of him from his books and articles, for the man has been largely lost from the historical record. Since it is impractical to organise this book around Lamarck's biography, I have arranged the chapters thematically, following as closely as possible the order in which these themes appeared in his writings.

2 Classification and scientific method

In the early part of his career, Lamarck was caught up in the excitements and frustrations of natural history. He came to the study of plants, animals and clouds with enormous enthusiasm, but quickly realised that translating the fruits of his observations into a coherent form posed a number of philosophical problems. Eighteenth-century natural history cast its net wide: all natural objects came within its scope, and they were to be observed, described, collected and catalogued. These tasks were the basis of classification. The broad scope of eighteenth-century approaches was not new, yet the passion for organising natural objects was exceptionally well-developed in this period. Naturalists such as John Ray (1627/8–1705) made more rigorous attempts at classification than the miscellaneous approaches adopted by earlier writers, who had often included monsters and mythical beasts in their schemes.

Classification raised questions which were at once metaphysical and methodological. The complexity of the philosophical and theological issues it prompted is indicated by the variety of systems and opinions current in Lamarck's time. One fundamental question was whether human beings were in fact capable of capturing the real relationships between different parts of nature by means of a taxonomic system. How *could* the human mind have knowledge of the physical world? If a genuine system could not be found, natural historians would have to remain content to name specimens according to admittedly artificial criteria for the sake of practical convenience—a situation which appeared

Lamarck

to do less than justice to the admirable works of the Creator. Indeed, how could God's creation be comprehended by human endeavour?

Underlying these debates were also questions about how nature really worked. Were there any divisions in nature or were they in the minds of men? Could a workable and accurate classification be produced at all? Natural historians agreed that there were observable relationships between natural objects, and also that a principle upon which they could be organised into smaller groups would be invaluable. There was no consensus however on what the nature of those relationships was nor about the organising principle required.

Before Lamarck's impressive contribution to the field in the *French Flora* (1779), the main actors in this intellectual drama were the Swedish naturalist Carl Linnaeus (1707–78) and the French philosopher and naturalist Georges, comte de Buffon, Lamarck's patron. The former, an orthodox Christian, had developed a confessedly artificial approach to the classification of plants—the sexual system. ('System' was the term used to describe artificial approaches, while those which attempted to be 'natural' were called 'methods'.) Linnaeus took the reproductive parts and arranged species according to the number, shape, proportion and position of the stamens and pistils (the 'male' and 'female' organs respectively). He justified using the generative organs by stressing their functional importance—they expressed the essence of the plant. Linnaeus was certainly aware that his system had shortcomings. He also came to recognise that the strict theological view that species were distinct and immutable was in some respects problematic. But in general he expressed confidence in the reality and permanence of species.

Linnaeus's most important contribution to botany was the development of binomial nomenclature. The practice of

Classification and scientific method

giving each plant two names, the first denoting the genus to which it belongs, the second its species, is still in use today. Linnaeus saw botany as an exercise in naming and ordering rather than in elucidating the anatomy and physiology of plants; it was a discipline for which the tools of logic were therefore of the utmost importance.

Buffon, a follower of Newton, was of a more sceptical cast of mind. He passionately rejected the rigid approach of Linnaeus, on the grounds that there are only individuals in nature, which form an unbroken continuum not divisible into unequivocal groups. Rather than imprison nature within an artificial and arbitrary system, he asserted that natural history must follow nature. The basic concepts employed by taxonomists, such as family and genus, were for him mere human inventions. Buffon believed that all human knowledge pertained to relationships, not to essences as Linnaeus had held. Thus in his monumental *Natural History* (*Histoire naturelle*, 1749–77) he arranged animals in the most 'natural' order available, according to their usefulness and interest to man; he started with domestic animals, with which human beings have an extremely intimate relationship, and gradually moved away to those with the most distant connection. Buffon's primary concern was with the real similarities between organisms, which could only be grasped if the entire animal or plant is considered. Nature, being subject to change and producing objects with fine gradations between them, cannot be understood in terms of logical abstractions. Buffon thus saw things in a quite different perspective from Linnaeus with his systematic approach.

France could boast of a distinguished botanical tradition, particularly associated with the Jardin du Roi. Joseph de Tournefort (1656–1708), Michel Adanson (1726–1806) and members of the de Jussieu family had laid the foundations for a natural classification through using as many characters

Lamarck

as possible in determining species and genera, rather than restricting themselves to the sexual parts as Linnaeus had. These men were Lamarck's intellectual ancestors.

All the branches of natural history employed the idea of a hierarchy of forms from the most rudimentary to the most elaborate, a kind of unbroken scale or chain of natural forms—an idea which derived from classical antiquity. This 'great chain of being' was constantly discussed by eighteenth-century writers. For them it was a powerful and convenient metaphor linking the whole of creation in a single, continuous series which stretched, through imperceptible links, from brute matter, via plants, animals, man, cherubim and angels, to God. In the second half of the century, naturalists increasingly ignored the supernatural part of the series above man, but the ideas of hierarchy and of continuity between forms associated with the great chain of being retained considerable influence in the study of the living realm.

We can interpret Lamarck's work as an exploration of this more restricted chain of organic beings since many of his key concepts derived from it. He did not, however, embrace all its aspects. He rejected the suggestion that there was a single series encompassing the whole of nature because he found animals, plants and minerals to be profoundly distinct. Furthermore, the very notion of a full chain of being had metaphysical implications which he could not accept. Most objectionable was the idea that there was a continuity between natural objects which could not be empirically examined because it pertained to the spiritual realm. Lamarck saw living nature as a number of different, unconnected series, admitting later in life that among animals there may be even more of these than he had originally imagined. His model of the organic world was not a single chain but a structure made up of several branches.

Eighteenth-century debates on classification and on the

Classification and scientific method

chain of being shaped Lamarck's earliest scientific work. He had to settle for himself what theological and methodological stance to adopt. From a religious point of view, Lamarck is best characterised as a Deist. God existed for him, in some remote sense, as the Creator, but nature functioned according to laws the explanation of which was entirely separate from all theology. He thus had no commitment to viewing organisms as the direct creations of God.

In terms of scientific method Lamarck's approach was 'naturalistic'. In other words, he sought to understand nature in its own terms without reference to metaphysics of any kind. It was classification, and more particularly the problems of plant taxonomy, that first led Lamarck to reflect upon the ways in which the human mind could have knowledge of nature and upon the processes of nature itself. As later chapters show, these early concerns of Lamarck played a key role in the development of his theory of transformism.

In his classificatory ideas, Lamarck took something from the perspectives of both Linnaeus and Buffon to create a more satisfactory instrument for natural historians. From Linnaeus he adopted binomial nomenclature and the use of the reproductive parts. From Buffon, to whom he was indebted in a myriad of ways, he adopted the idea that nature was full of change, and a healthy scepticism about the value of abstract, arbitrary systems based on logic rather than on observation. Natural science should, Lamarck thought, be built on the foundation provided by recording the effects of natural laws in observable physical phenomena. This was the heart of his naturalistic method which, like Buffon's, stressed the importance of understanding the relationships between different parts of nature. This repudiation of metaphysics in favour of knowledge acquired through the senses is important because it is indicative of a

Lamarck

more general shift in scientific thought in which the limitations to knowledge imposed by the nature of the human mind were increasingly stressed.

In the *French Flora*, Lamarck argued that there were two distinct goals in botany which must not be confused. The first was to construct a simple, quick and reliable route to the identification of particular specimens: this entailed making distinctions not sanctioned by nature. The second was to grasp a conception of the entire plant kingdom as an ensemble, and to find the means for expressing these complex relationships by revealing the various levels of structural complexity within the graduated chain of plants. These tasks were analytically distinct, that is, they were two different ways of organising a section of nature. Lamarck set out to accomplish both of them.

The two procedures these distinct tasks entailed were set out separately in Lamarck's publications. The first procedure made use of a table, an 'analytical key', to help the reader identify the specimen by leading him to the genus to which it belonged. In the key, each genus was given a number which referred to the second part of the book. There the genera and their constituent species were described, the habitat indicated, and all the parts of the plant were mentioned. The key, by contrast, only used the reproductive parts.

The second goal, understanding the relationships between plants, was, of course, far harder to attain than the first. Although a single part or group of parts such as flowers, stamens or sepals, was sufficient for the purposes of identification, the whole plant now had to be considered, in order to reveal its similarities and dissimilarities with other plants. The naturalist who strove to reflect nature in his classification needed to have a sense of each species as a totality; he could then move on to group plants together in genera, families, orders and classes. Thus, the Brussels sprout plant

Classification and scientific method

and the cauliflower belong to the same genus (*Brassica*), a member of the *Cruciferae* family, which also includes water-cress, wallflowers and stocks. *Cruciferae* are part of the larger category *Rhoeadales* (the group whose type is *Rhoeas*, the poppy). Finally, according to modern taxonomy, they are part of the *Spermatophyta*, that is, seed-bearing plants, a division of the very greatest generality.

Lamarck admitted the practicality of Linnaeus's sexual system, as his use of it in his analytical key indicated. But he was critical of it on the grounds that using a single part inevitably led to distortions. His reservations about the sexual system are important. In the *French Flora*, he stressed the need to comprehend the plant as an integrated organism, and suggested ways in which the hierarchy of organisational complexity could be expressed in a natural order. In his botanical work he tried to reconcile a natural order with a workable classification. His commitment to natural classification as a long-term goal, despite his scepticism about the tendency of human beings to impose arbitrary categories upon nature, informs most of Lamarck's scientific work. It also constitutes a fundamental link between his botanical and his zoological researches.

Lamarck's prerequisites for identifying a plant were that the vocabulary used to describe its anatomy be unequivocal, that the specimen be observed accurately, and that no attempt be made to classify cultivated plants by the same criteria as wild ones (the *French Flora* dealt only with the latter). His analytical key consisted of a series of dichotomies in which there was a choice at each stage between two mutually exclusive features, such as between bisexual and unisexual flowers, or between flowers with and without petals. By this process of elimination, plants could be identified quickly, easily, and with great accuracy. The bases for discrimination were the reproductive parts: flowers, ovaries, stamens, calyx and corolla. He suggested that there

Lamarck

should be as many choices as possible, since this enabled one to reach the species level most rapidly. Lamarck called this process 'analysis', and used it, in various forms, in many of his books. 'Analysis is the name we have given in Botany to the method of dissection, by means of which one goes from the totality of all known plants down to each of them in particular; having only, at each point, to choose between two mutually exclusive characters' (E i 42).

For Lamarck, classes, orders, families and genera were human constructions having no basis in nature; only species were, in this sense, real. The others were quite simply categories imposed upon nature by the human mind as a means of comprehending the extraordinary diversity of living things. But Lamarck did *not* conclude from this that the naturalist was entitled to define and apply these terms in an arbitrary way. On the contrary, the major divisions of the plant kingdom should approximate as closely as possible to whatever natural groupings do exist. For example, Lamarck criticised the division of plants into trees, shrubs and herbs, on the grounds that genera based on functionally significant characters (as opposed to trivial ones like flower colour) often contained species from all three groups. He concluded that the division was more artificial than was necessary.

Since classes, families and genera were groupings at different levels of generality, they were defined by different kinds of criteria. Classes, being the most inclusive category, had to be defined broadly enough to include a wide variety of forms, while genus was a more restricted division designed to incorporate fewer members. For this reason classes, families and genera could, to a certain extent, be treated separately. Each head was defined by criteria which corresponded to its level of generality. Lamarck applied the same approach to dividing up the animal kingdom. Thus mammals are a class for Lamarck, defined as 'viviparous

Classification and scientific method

with mammary glands, two or four articulated limbs, respiration through lungs ... hair on some part of the body' (p 1.280). In their turn mammals were made up of four orders: whales, seals, hoofed mammals and unguiculates. This last order was defined in quite specific terms, 'four limbs, flat or pointed nails at the end of their fingers which do not cover them'. Eight families made up the order, the last being the quadrumanes, which included such genera as baboons and orang-utans (p 1.344–7).

In his botanical dictionary, Lamarck used the families delineated by his colleague at the Jardin du Roi, A. L. de Jussieu (1748–1836), but not his genera. The genus was, in fact, a particularly important grouping because it provided the first half of a plant's name, and Lamarck wished it to be named so as to evoke a character shared by all its members. For example, the plant genus *Sagittaria* (from Latin *sagitta*, 'arrow') is made up of plants with arrow-shaped leaves. Furthermore, the character so used should, ideally, also express the level of complexity of the genus in the plant kingdom as a whole.

In the analytical table Lamarck used different classes from those in the descriptive section. It was in the latter that it was imperative for the major groupings to be as natural as possible. In 1806, as a result of Lamarck's collaboration with the Swiss botanist Augustin-Pyramus de Candolle (1778–1841) there appeared the *Synopsis of Plants described in the French Flora* (*Synopsis plantarum in Flora Gallica*)—Lamarck's only Latin publication. In contradistinction to the first section which used five classes based mainly on the reproductive parts, its descriptive section used cotyledons (rudimentary, embryonic leaves) as the basis for forming the classes. Plants were divided into those with no cotyledons, one cotyledon and two cotyledons. Their use in the *Synopsis* may indicate a growing commitment to physiological and developmental features as

Lamarck

classificatory criteria. This was a shift away from the formal approach of earlier taxonomists, such as Linnaeus, with their emphasis on external characters, and it was clearly visible in Lamarck's own changing ideas about classification.

Lamarck recommended three procedures for determining the natural order of the plant kingdom. First, fixed points in the series had to be established by deciding which were the most and the least 'perfect' plants. By 'perfect', Lamarck meant complex or structurally elaborate, and his choice of terminology is revealing in that the idea of perfection implies a fixed, independent standard by which each species may be judged. This was an implication Lamarck was not anxious to accept, yet he continued to use the word. However problematic the ideas underlying them, the fixed points acted for Lamarck as bases around which the rest of the series could be built. He spoke of degrees of organisation (that is, levels of structural complexity), and implied that elaborate forms were more 'alive' than their simpler counterparts. Lamarck employed concepts like natural hierarchy, levels of vitality and continuity between members of series which were to play a central role in his later work, reminding us yet again of the number of ways in which the technical problems of plant taxonomy spawned the biological inquiries of his later years.

In the *French Flora*, Lamarck placed the most complex plant in first place and worked down from there. There was, of course, no hint of a historical progression of species at this point. Lamarck's second means of determining the natural order of plants was to set up rules for finding relationships between species. Although it was preferable to take the plant as an ensemble, he claimed that individual parts were useful in determining relationships in direct proportion to their prevalence in the plant kingdom. A part all plants possessed (such as the seeds) was of more value than one found only in a small number. He then established

Classification and scientific method

a table of numerical values indicating the universality of the part in question. Lamarck's third procedure consisted of the use of the overall degree of perfection to establish the natural order where there were no other clear lines of demarcation.

The project of writing a new flora for French wild plants introduced Lamarck to a number of complex scientific issues, which were basically methodological and which were not specific to botany but concerned the natural sciences in general. The problems he tackled in his early work were broad in scope, and we can trace their development throughout Lamarck's later work.

Botany was devoted to the study of only one part of nature. In the *French Flora*, as in his other works, Lamarck set his detailed natural history in a broad context. He linked plants and animals together as possessing a 'vital principle'— a concept he later abandoned; he placed the world of inorganic nature in a quite separate category. Animals differed from plants, he argued, since they displayed 'sensibility', that is, the capacity to have sensations, and spontaneous movements, both of which plants lacked. He went on to emphasise the gap between man and the animals: the former, having the gift of reason, was closer to God. The manner in which Lamarck developed these definitions shows at once the impact of his early reflections on the classification of plants on his subsequent work, and his departure, in the course of a long career, from some of the views he held in his youth. He retained his early preoccupation with levels of organisation throughout his life, and also his conviction that the natural world was composed of organic and inorganic parts, which were fundamentally different. The criteria by which he separated them changed, however. Lamarck always maintained that there was no simple continuity between plants and animals, although he ceased to use sensibility as the differentiating

Lamarck

factor, and came to use irritability, the ability to contract when touched, a property he thought only animal tissues displayed. It was in his approach to man that Lamarck underwent the most marked change of mind. He closed the gap between animals and human beings, coming to view the difference as residing in degree of complexity and not in a unique faculty of human reason.

Classification took on new significance with the development of Lamarck's theory of transformism. Previously, natural order had referred to static patterns of relatedness, but it now came to mean the sequence in which natural forms had actually been produced. Lamarck reached the conclusion that levels of increasing complexity and degrees of kinship were the results of real historical changes.

The problems of taxonomy formed only one aspect of the discussion on method found throughout Lamarck's writings. Equally important were questions of the origins of human knowledge and the role of language in thought. Here Lamarck was clearly indebted to the French philosopher l abbé Condillac (1714–80). Following John Locke (1632–1704), Condillac rejected the notion that man's understanding works with innate ideas, proposing instead an empirical approach to the workings of the mind which rigorously traced the origins of ideas from sensations. All knowledge therefore had its foundations in experience. Condillac called his method 'analysis'.

Locke had argued that the complex ideas of the human understanding could be broken down into their simple constituent sensations and impressions. What Condillac added was a theory of language which demonstrated that symbols were necessary for thought to take place—that the experience of the senses passed through language to become thought. Indeed, for Condillac and his followers (the *idéologues*, students of the science of ideas), the natural

Classification and scientific method

sciences could only make progress when they employed the correct language. This conviction was of some importance for classification. For both identification and finding natural order, naming was important. Although the difficulty of capturing the complexity of nature in language was recognised, giving appropriate names to natural forms was seen as a decisive step towards understanding them and communicating that insight to others. Constructing an appropriate language was therefore integral to the pursuit of science.

While Lamarck did not agree with Condillac on all points, the latter's emphasis on the way the mind operates, on the analytic method and the study of origins, and on the general importance of language, deeply affected Lamarck's whole outlook on science and scientific method. In botanical classification Lamarck was concerned with the relationship between names, ideas and the objects they represent. The function of language, in this case the names of plants and their parts, was to aid thought and memory. Lamarck wished to see a stable biological language develop, avoiding the arbitrary changes in nomenclature which occurred when people named for the sake of naming, so that established patterns of thought could progress rather than be wantonly disrupted. At the same time, the language had to be flexible enough to allow new names to be added without causing chaos. Signs were most effective as tags for ideas when they expressed as accurately as possible the nature of the object they designated and its relationship to other objects.

Influenced by Locke and Condillac, Lamarck, like so many of his contemporaries, held the view that all knowledge came ultimately from the senses. It was a logical consequence of that belief that scientific accuracy depended on the quality of the observations made. People had to be trained to use their senses in particular, and their minds in

Lamarck

general, more effectively, for it was readily acknowledged that no other guarantee of correct knowledge existed. Since education and life-experience affected mental processes, people perceived the world in very different ways. The emphasis given by Lamarck and his contemporaries to training scientists in the skills of observation was one means of overcoming the relativism implicit in their position.

The problem of finding secure foundations for scientific knowledge was compounded for Lamarck by the workings of nature as well as by the character of the human mind. He believed that nature did everything so slowly that it was hard to perceive changes except over very long periods of time. Compared with the speed of natural processes, the human lifespan was rather short. Earlier writers had drawn attention to the distorted vision which could result; just as rose bushes could assume gardeners to be immortal, naturalists believed the work did not change because they had not seen it do so. According to Lamarck, it was imperative that models and analogies be employed allowing gradual changes to be explained along the lines of faster ones which were easier to observe. This method was strengthened by Lamarck's conviction that nature was economical, using similar laws and mechanisms in all its works. In order to 'see' nature in transformation—to construct nature's progressive operations—the appropriate framework was needed to render that knowledge compatible with human psychology. Science and epistemology went hand in hand. Understanding how the human mind worked assumed a central role, not just as a branch of the natural sciences, but as the sole proper basis of scientific method.

3 Plants

Lamarck's interest in botany went beyond the taxonomic problems that the discipline raised. He continued to write about plants throughout his career, and discussion of their definitive features is found in most of his published works. In the 1780s and 1790s he became increasingly preoccupied with defining the properties common to all living things. The comparison of plants and animals displayed, for him, the basic powers of nature to produce living, organised beings. Plants and animals, properly apprehended through an appropriate classification, showed their differences to be differences in vitality. In other words, life could be present in a body to a greater or lesser degree, and studying the various forms of life was the central task of biology. By looking at Lamarck's understanding first of plants and then of animals, we can come to appreciate his interpretation of the nature of life and its position as the central concept underlying his biology. In fact, his biological work encouraged him to refine and develop his methodological principles.

Given the historical context, it is clear that Lamarck had every reason to be attracted to the study of plants. Botany occupied a special place in eighteenth-century culture. Collecting and studying plants was all the rage, offering a unique combination of aesthetic and intellectual delights. Elaborate arguments about classification were waged largely in relation to botany. But the general problems of putting nature's manifold productions into a coherent framework applied equally to animals and minerals. Linnaeus, for example, was interested in all three realms, and attempted the classification of each one. Lamarck too saw the questions he had tackled in the *French Flora* as directly

Lamarck

comparable to those raised by zoology and mineralogy. As he was well aware, the plant kingdom, like other parts of nature, could be approached in a number of different ways. He had strong views on what the priorities of the science as a whole should be; taxonomy was merely one aspect of it. As a science, botany was not, he believed, confined to the identification of specimens and the reconstruction of natural affinities between plants; it should also investigate the nature of plant life—'the physics of plants', as he called it.

In the *French Flora* he had already hinted at his fascination with the phenomenon of life which animals and plants shared. In his later writings he had much to say about the distinctive contribution botany could make to the new science of biology. As he spent more time thinking about organic nature, he used plants in two ways: to find out what characteristics all living things shared by investigating the common features of plants and animals, and to clarify plant and animal differences. Lamarck had a broad vision of the job a natural historian should do. It made no sense to him to treat one part of nature in isolation from the rest: he had a view of nature as a coherent whole. He possessed faith in the value of correct definitions of natural categories such as 'plant' and 'animal' as a solid basis for detailed scientific work.

His notion of the definitive characteristics of plants is of some importance for the development of his biology. Plants, Lamarck asserted, possessed merely the faculties which were essential for life to exist. They fed, grew, reproduced and made the substance of their bodies from nutriments; they transpired and absorbed particles from their environment. Like animals, plants manifested states of health and disease, and, since they were alive, eventually they had to die. They therefore played a useful analytical role in Lamarck's project of developing a science of living things by showing life in its unadorned state. Like the

Plants

invertebrates he spent so long studying, plants showed nature's operations in one of their simplest forms and thereby shed light on the fundamental properties shared by the whole organic realm.

Plants and animals, according to Lamarck, shared other features in addition to those which were necessary for life to exist. Two analogies in particular caught his attention. First, both realms could suspend their active life by 'hibernating' when it was excessively cold. Lamarck employed the analogy as an explanation for the sudden bursting into life of apparently dead plants in the spring. During this suspension, the 'vital knot' at the base of the stem and top of the roots maintained the plant in a living state. To Lamarck this illustrated an important difference between plants and animals. Plants were clearly less vital, since they could survive with only one part in a state of vitality, while all the parts of animals are alive.

The second analogy was that both kingdoms displayed communal forms of life—many individuals living together in a colony. Lamarck saw trees, for example, as made up of small individuals such as leaves, which live only for a year, just reproducing and then dying, while the stem or trunk persisted and did not age in the way most other living things did. The possibility of grafting appeared to be supporting evidence for the colonial character of trees.

Lamarck envisaged a formal parallelism between plants and animals in their taxonomic groupings. Under the term 'class' in his *Botanical Dictionary* (*Dictionnaire de botanique*), published between 1783 and 1795 as part of the *Encyclopédie Méthodique*, Lamarck included a table which showed the main classes of animals and plants in two corresponding columns. Each realm was divided into six classes. The table suggested that the simple forms of plants and animals resembled each other more than the complex ones. In 1786, Lamarck presented the comparison as a

Lamarck

pleasing pattern, but in the light of transformism it can be seen as an abstract of two simultaneous historical processes, with polyps being the 'rough draft' of animals as cryptogams (ferns, mosses, fungi, etc.) were of plants.

Despite his growing commitment to the study of the general properties of organisms, Lamarck was struck by the differences between animals and plants. He stressed that there was no continuous chain linking all organic beings: there were no nuances between plants and animals, rather a 'clear-cut line of demarcation'. He repeatedly and emphatically rejected as figments of the imagination the postulated intermediates, such as the zoophytes (literally, animal-plants) so beloved of eighteenth-century naturalists. He also found plants the less perfect of the two groups. They were passive and stationary, while animals were active and mobile. They reacted to stimuli only slowly, if at all. They had no differentiated internal organs. Plants and animals were chemically distinct, with, so Lamarck thought, carbon predominating in plants and nitrogen in animals.

> Nature began the production of animals and plants at the same time, starting work on bodies which were essentially different by virtue of their chemical elements. Everything which nature managed to produce in one group was different from that which she was able to produce in the other, although she worked in both cases to extremely analogous plans. (A i 83)

Lamarck rejected the idea that 'sensitive plants', such as the famous *Mimosa pudica*, whose leaves responded to touch, displayed animal-like capacities. He explained their reaction purely in terms of hydraulics—the build-up of fluids inside them. They did not thereby show the capacity to be 'irritable' or genuinely 'sensitive', both of which he saw as the prerogative of animals.

The terms irritability and sensibility and their cognates

Plants

were commonplace in eighteenth-century natural history and medicine. Frequently they were understood as organic properties, that is, as unique to living things. But there were no definitions of irritability and sensibility which commanded general assent; they could be used in different ways according to the distinctions which were deemed most significant, be they between living and inert, between animal and plants or between consciousness and its absence. One of the most influential formulations was that of the Swiss physician Albrecht von Haller (1708–77), who associated irritability with muscles and sensibility with nerves, thereby identifying irritability with the responsiveness of living matter to stimuli and sensibility with perception and consciousness. Haller located these properties in specific tissues—muscle and nerve respectively. The association of organic properties with specific parts opened up the possibility of a physiological rather than a mechanical exploration of the relationship between structure and function. Haller's initially simple distinction offered a framework upon which a language could be constructed for the exploration of the structural complexities of living things. Lamarck's use of this vocabulary, together with his changing definitions of the terms 'irritability' and 'sensibility' must be understood in this context.

In the *French Flora*, Lamarck had attributed sensibility to all animals to distinguish them from plants. In his subsequent work he developed a more refined approach to differentiating plants from animals, and different classes of animals from each other. He came to believe that it was in fact irritability which separated animals from plants. Far from being a general animal property, sensibility was, Lamarck claimed, restricted to certain groups of animals.

> *Irritability* is the most general animal character ... All plants ... are completely lacking in *irritability* ... Its

effect consists of a contraction which any irritable part undergoes the moment it is in contact with a foreign body. The contraction ceases when its cause does, and is repeated as many times after the relaxation of the part as new stimuli irritate it. (p i 93)

The so-called sensitive plant failed to fulfil these criteria of irritability. Far from being present in plants, sensitivity was restricted, even among animals, to those groups which had the appropriate organs. With no differentiated internal organs, plants could only possess those limited organic properties which were necessary for life to exist, and which by definition did not require specialised structures. This was Lamarck's way of enunciating a fundamental principle about the tight linkage between structure and function upon which his zoological work depended.

Lamarck was interested in the special qualities of living systems and he sought to distinguish precisely the properties all organisms shared from those that only some possessed. In the mechanistic tradition of Descartes, which had been influential since the mid-seventeenth century, organisms were simply complex machines, and an organ a place where a function occurred. For Lamarck, on the other hand, the organ and its function had developed together. In the first case there was only a loose connection between structure and function, in the second there was an inseparable one. In Lamarck's view, biological capacities arose out of the intimate, dialectical relationship between organic functions and the structural organisation of tissues. This particular way of understanding the relationship between structure and function played a fundamental role in Lamarck's thought on the order and manner in which organic forms arose.

The biological inferiority of plants compared with animals was manifested in the greater dependence of the former

Plants

upon the environment; unlike animals, plants were moulded by their milieu in a direct way. Nourished by fine fluids in the atmosphere (see p. 49) and by liquids absorbed from their surroundings, and unable to move for food or reproductive purposes, plants were of necessity bound to their habitat. When he elaborated the major statement of his transformist position, the *Zoological Philosophy* of 1809, Lamarck used botanical examples as convincing evidence of the delicate balance between organism and environment. Plants offered an impressive range of examples to illustrate life's dependence on the environment and the plasticity of organisms in accommodating environmental changes. As he recognised, such changes also resulted from human action: the cultivation of new varieties of plants for food and hybridisation.

At the same time plants, with their lesser vitality, were unsuitable material for exploring the more complex action of organic phenomena of a higher order. In biology the study of organisms with few internal organs had to be enriched by study of those with extremely elaborate organ systems at the other end of the scale of complexity. Lamarck's growing concern with the nature of life could be taken further only by zoological research. Classification and the study of natural relationships, important and challenging as they undoubtedly were, had to be supplemented by physiological, psychological, physical and chemical methods if the new discipline of biology was to find adequate expression as a science of life.

Ultimately, plants presented Lamarck with less of an intellectual challenge than animals because, despite his rejection of a single chain of being, he did accept a hierarchy of vitality. In that series animals were unquestionably superior to their more mechanical vegetable cousins. To move towards an understanding of his notion of biology,

Lamarck

therefore, we have to consider how Lamarck conceptualised the whole range of vital expression, and this leads us to his study of animals.

4 Animals

When, with the reorganisation of the Jardin du Roi as the Muséum National d'Histoire Naturelle, Lamarck became Professor of Insects and Worms, he was no stranger to zoology. Writing on botany in the 1770s and 1780s, he had brought to the study of plants an awareness of the issues currently being discussed in natural history, medicine, and natural philosophy. He combined an interest in the general properties of plants, animals and minerals with detailed taxonomic work. Lamarck was an expert in the field of shells, having himself built up a vast collection. He was particularly interested in fossil shells, sharing the growing appreciation of the crucial role fossils could play in reconstructing the history of the earth, its flora and fauna. Like Buffon, Lamarck saw petrified animal and plant remains as 'precious monuments' of nature's past. But unlike his colleague at the Muséum, Georges Cuvier, who was fast becoming the leading authority on the interpretation of fossils, Lamarck refused to accept that the history of nature had been punctuated by large-scale catastrophes (such as a flood) which had permanently destroyed many species at a stroke. Cuvier suggested that fossils were the remains of organisms killed by geological cataclysms. He also adhered to the view that species were created, although not on only one occasion. Lamarck rejected both the extinction and the special creation of species. Sudden revolutions in the natural order were inherently implausible from his point of view, which rested on a belief that the laws of nature acted slowly, regularly and gradually, maintaining nature in a state of equilibrium. He assumed that the past was like the present, and that it should

Lamarck

therefore be explained in terms of causes currently operating, not ones which had never been observed. This position—described as 'uniformitarian'—was held by the Scottish natural philosopher James Hutton (1726–97) and was later associated with the important geological work of Charles Lyell (1797–1875), Darwin's mentor. Integral to uniformitarian ideas was the belief that nature was a balanced system, resulting from slow, piecemeal alterations. This did not *necessarily* imply an evolutionary view of the history of organic life, although for Lamarck they did indeed go together.

In working on fossil shells, Lamarck was particularly interested to see how far fossils and living species were similar. By 1799 to 1800, when his transformist views were crystallising, Lamarck claimed that a low degree of similarity was sometimes to be expected. Fossils differed from living forms because species, being mutable, had, over thousands of years, changed into those we now know. Those who used catastrophes to explain geological change accepted that there was a succession of fossil forms in sequences of geological strata. But this could be construed as evidence for extinction followed by the creation of new species. By contrast, Lamarck argued that new species, particularly of marine invertebrates, were still being discovered. Thus categorical assertions about extinction resulting from catastrophes were, for Lamarck, premature and unscientific when knowledge of the natural world was still incomplete. Geological catastrophes and extinction were seen by Lamarck as alternative explanations to his own; accepting that either had occurred would therefore have entailed abandoning his theory of the transformation of species.

Lamarck's work on conchology was published in its fullest form in his *Natural History of Invertebrates* (1815–22), and was, so admiring contemporaries claimed, 'universally adopted among Naturalists'. Several editions of

Animals

plates were published to match the descriptions and classifications. The evident success of Lamarck's contributions to invertebrate taxonomy and palaeontology is important, but should not be overestimated. Having made the points that extinction was an untenable hypothesis, and that fossils differed from living forms because of the twin effects of organic transformations and environmental change, Lamarck paid surprisingly little further attention to shells or to fossils in his general scientific writings. They were certainly an important element in the *genesis* of his historical approach to nature, but a detailed examination of the fossil record had no place in his arguments for transformism.

The two opposing positions on transformism and extinction were demonstrated in the scientific debates concerning mummified animals brought back from Napoleon's expedition to Egypt between 1798 and 1801. Upon examination it was discovered that these animals were morphologically identical to living specimens. Lamark claimed that this was precisely as he had expected, considering that environmental conditions had remained constant, and that four thousand years was a short time relative to the history of the earth. Others, who were pressing forward the case for extinction against transformism, drew the opposite conclusion. No changes had taken place in four thousand years because species were fixed types, incapable of the kinds of transformations Lamarck had in mind; it followed that animals were liable to become extinct should their environment alter. Lamarck's colleagues at the Muséum, particularly Cuvier, believed the mummies were a crucial test case that would prove finally whether transformations or extinction, which they too perceived as alternatives, occurred. Lamarck remained undaunted by their criticism of his position. His detailed and impressive work on conchology convinced him of the possibility of transformism and thus

Lamarck

assisted the development of his mature biological philosophy, but once this took on a more rounded shape, palaeontology played little part in the articulation of his science of life.

Lamarck's first sustained and important discussion of the nature of animal life was in his *System of Invertebrate Animals* of 1800, a largely taxonomic work to which he added the opening lecture of his course that year at the Muséum. Here, as elsewhere, Lamarck approached his materials as if for the first time, taking immense care to set out his basic principles and define his terms afresh. How animals should be classified was an enduring concern, which was not diminished by his transformism, and he continued to make significant modifications to his analysis to the end of his life. An effective classification took account of the nature and history of living things for which proper definitions of life, matter and motion were indispensable. Such definitions were inseparable from Lamarck's basic methodological assumptions as to how the organic realm may best be studied. His starting-point was always nature as a whole, which was then systematically broken down into its constituents, using an approach reminiscent of the analysis he had applied to plants. He began with the division of nature into inert and living; the latter was then separated into plants and animals. Animals in their turn were composed of invertebrates and vertebrates, each of which were resolved into classes, orders and so on.

The famous opening lecture delivered in 1800 contained the first detailed published statement of Lamarck's transformism. Having first established the fundamental division between animals with and without backbones, he divided invertebrates into seven classes (molluscs, crustaceans, arachnids, insects, worms, radiates and polyps). Here he presented the animal series in order of *decreasing* complexity, discuss-

Animals

ing the four classes of vertebrates (mammals, birds, reptiles and fish) before moving on to the invertebrates. He did not start, as he claimed nature had, with rudimentary forms, but with the most elaborate animals, to show the 'degradations' of form over the animal series:

> For [invertebrate animals] show us still better than the others that astounding degradation in organisation, and that progressive diminution in animal faculties which must greatly interest the philosophical Naturalist. Finally they take us gradually to the ultimate stage of animalisation, that is to say to the most imperfect animals, the most simply organised, those indeed which are hardly to be suspected of animality. These are, perhaps, the ones with which nature began, while it formed all the others with the help of much time and of favourable circumstances. (N 237)

Although his later *Zoological Philosophy* presented animals in the actual order nature had taken to produce them, an important element from his earlier approach remained. Simple animals were seen mainly in terms of negative characteristics: to different degrees they lacked the faculties higher animals possessed. The most advanced animal forms were the standard with which all others were compared. For example he described infusoria and polyps, the simplest animals, as having 'no nerves, no vessels, no other internal organ specially for digestion' (P i 277).

The simultaneous emphasis on the most perfect and on the most rudimentary animals enabled Lamarck to establish, as he had done earlier for plants, the two extremes of a natural series: the minimum and maximum degrees of animality. The space between the end points could then be filled in. The criteria for dividing up the invertebrates were the respiratory and circulatory systems and the nervous system, while for vertebrates the reproductive organs, as

Lamarck

well as those of respiration and circulation, were used. We can see here once again the *idéologue* tradition of analysis (see p. 22), in that Lamarck began with the more complete animals, those with the most highly developed and differentiated vitality, and searched for their more primitive and basic expressions in progressively less developed animals. He discovered the building blocks of nature by moving down the animal series; he then presented the animal world as nature's sequential elaboration of higher forms from the basic elements of life.

The details of Lamarck's 1800 lecture were spelt out in the main part of the *System of Invertebrate Animals*. Invertebrates were presented as a single series of classes characterised by the lack of a vertebral column and jointed skeleton. Analytical tables, as in the *French Flora*, were used for each class in the *System* to break it down into numbered genera. By 1809, Lamarck had added three new classes to the invertebrates—infusorians, annelids and cirripedes. These are in order of *increasing* complexity. In the *Zoological Philosophy*, Lamarck presented a 'table of the distribution and classification of animals, following the order most like that of nature' (P i 277–80). A further significant innovation there was the grouping of classes into six degrees of organisation, four among the invertebrates, two among the vertebrates. The descriptions of both the degrees of organisation and the classes emphasised the negative characters of simple animals and the positive characters of complex ones.

Lamarck drew extensively on the nervous and digestive systems to discriminate between classes. The use of the former is particularly significant, for it shows Lamarck's concern with animal behaviour and its physiological basis, the nervous system, as the most significant manifestation of the special properties of life. His preoccupation with the nervous system was even more pronounced by 1815 when,

Animals

in the *Natural History of Invertebrates* all animals were grouped under three heads: apathetic (that is, not reactive), sensitive, and intelligent, thereby bringing behaviour to the fore as the most general classificatory criterion. Only vertebrates qualified as 'intelligent', for their nervous systems were sufficiently elaborate for them to have ideas which they manipulated. 'Sensitive' creatures comprised the six most complex classes of invertebrates: because they had the physiological capacity to receive sensations, but not to form ideas, they could not perform complex mental operations in their nervous systems. The four most primitive classes Lamarck designated 'apathetic' to convey their passivity and dependence on the environment; they exhibited only the definitive animal characteristic of being irritable.

Thus, as always, Lamarck was particularly involved with the problem of forming large, general groups, for these revealed the principal contours of the kingdom in question. In both his zoological and his botanical work, he recognised that different procedures were required in assigning organisms to their classes, genera, and species respectively. We can see the use of the irreducible properties of life to lay down the most general distinctions between what Lamarck called the 'principal masses', the large clumps of broadly similar animals. These 'masses' showed most clearly the overall progression of life across all the animal classes. To the extent animals were to be understood as forming a series, it was revealed by this rather than the specific level. The idea of a series was a valuable one for Lamarck, as it was for others concerned with classification, because it suggested an ordered pattern of natural relationships between members of a group, without restricting the precise form the relationships took. Series was, like so many of Lamarck's other basic concepts, of considerable heuristic value. The principal masses revealed the gradual increase in

Lamarck

the complexity of animal forms because they were relatively free from the 'anomalies' often produced by the environment, which distracted the naturalist from perceiving the true taxonomic location of a given species. When discussing rules for determining natural relationships between animals, Lamarck distinguished between those parts which had not been changed by external circumstances, and those which had, generally finding the former more taxonomically reliable.

Generally, Lamarck found organic parts which had been changed by accidental environmental factors to be unreliable for taxonomic purposes and best set aside in favour of nature's unblemished productions. He therefore rated internal organisation more highly than external appearance as a source of significant characters for classificatory purposes. External organs, those which were least essential for life, were most affected by the milieu, and consequently were more likely to display 'anomalies' or deviations from the animal series. This system of priorities which Lamarck employed differed markedly from the approaches of later evolutionists and geneticists who frequently used external, peripheral characters in experimental work on organic variations. For them, external features shed light on the whole evolutionary process, whereas for Lamarck they were the result merely of fortuitous environmental conditions. Lamarck's deep commitment to the idea of the overall harmony of nature made him less interested in the products of accidental circumstances.

Lamarck set out criteria for organising each level of the classificatory process. To determine the principal masses, the whole internal organisation was to be taken as an ensemble, while to uncover the relationships between species, definite morphological, and often external features were required. Having grouped animals into natural clusters or masses, the next step was to arrange them in the order of their physiological and anatomical complexity. The order was determined by reference to a fixed standard—the human body, the most complex animal form known by virtue of its highly developed brain and nervous system and the faculties these gave rise to, and its capacity to respond in an elaborate way to a changing environment. By 1815, however, since Lamarck's model of the animal kingdom included a number of series or branches, the use of the human being as the sole standard became problematic. Animals, he now believed, were built on different plans, so that the taxonomic criteria employed for each branch would inevitably differ somewhat. Priority was invariably given to organ systems which were either similar to or analogous with ones present in human beings. General organ systems which exhibited progressive variations (for example, respiration going from gills to lungs) were of greatest taxonomic value. These systems were, in order of decreasing importance: the organs of digestion, respiration, movement, generation, sensation (i.e. the nervous system) and circulation.

In practice, Lamarck accorded much greater significance to the nervous system and behaviour than the list above suggests because he was, above all, interested in the actions animals performed. The nervous system appeared the obvious key to this and he conceived of it as an index of vitality. The special emphasis on the nervous system is most marked in the *Natural History of Invertebrates* (1815–22). Another taxonomic precept reinforced this emphasis. When two versions of an organ system were compared, Lamarck asserted that the system showing the greater degree of similarity to a more complex form indicated more clearly the natural relationships. Lower animals were judged in terms of their absences, what they

Lamarck

lacked by comparison with vertebrates. No system symbolised the superiority of perfect animals better than the nervous system.

Although the underlying beliefs and principles regarding animal classification were relatively stable in Lamarck's mature biological work, he remained willing to contemplate modifications of details. In a supplement to the *Natural History of Invertebrates*, he added a new class, the ascidians, and remarked that he now saw two distinct series among invertebrates, with vertebrates forming a third. The nature of the links between these groups remained unclear, clouded by haphazard 'anomalies'. The 'march of nature', which Lamarck saw in the natural relationships between organisms, could not simply be inferred from close observation, because those living forms were the often irregular results of the continuous interaction between the 'power of life' and the randomness of the environment. Unravelling these two elements was a formidable task, but the naturalist could help by presenting the animal kingdom in two different ways: first, as a simple series showing the general progression towards higher forms, which would serve a principally pedagogic purpose; second, as a number of branching series depicting more faithfully the actual relationships within the animal kingdom. To the more fundamental conceptual difficulty—teasing apart the respective contributions of the laws of organic nature and of the environment—there was no simple solution.

Lamarck sought to capture the distinctive forms of life found among animals, or more precisely, to provide a scientific account of phenomena which earlier generations had been content either to attribute to vital principles or spirits or to reduce to matter and motion. Taking his cue from eighteenth-century physiology, he examined simple animals to discover how they functioned, and from there he arrived at a list of faculties common to all creatures.

Animals

Animals contracted when touched (i.e. they were irritable), they moved, they digested foreign substances, performed vital motions with the fluid and solid parts of their bodies, and, among the immense variety of animal forms, a few were even capable of abstract thought. When animals adapted to their surroundings their changes were active, not passive or instituted by God, as natural theology would have it. All these qualities stemmed from a 'power of life', which reached a peak at the top of the animal kingdom with mankind.

Lamarck's writings about animals, like every other branch of his work, reveal a commitment to unravelling the laws governing organic nature. These laws were to be discovered by the analysis of living beings, working from the complex to the simple, in the *idéologue* fashion. Elucidating the regularities of the living world could not be done by examining the properties of inert matter, for this, left to itself, had no special organic properties whatsoever. Organic nature emerged through the operation of laws of nature. Hence the fundamental divisions Lamarck saw between living and non-living matter, between plants and animals, between invertebrates and vertebrates, represented different stages in the operation of such laws. Lamarck aspired to the elucidation of the laws of living nature and to the demonstration of hierarchical levels of complexity which were the incontrovertible evidence of their action.

5 Life

Life was the central concept around which Lamarck's philosophy was arranged. He defined life as a physical phenomenon: this carried the important implication that it was perfectly amenable to scientific analysis, and that it was inappropriate to speak of a soul or a vital spirit; life was to be robbed of its mystery. Lamarck's approach to the nature of life rested on the conviction that, once the phenomenon had been correctly defined, the way was open to the new science of biology.

Lamarck defined life in physical terms, yet he also regarded it as a separate level of nature. To put it in more modern terms, Lamarck was against reductionism, the explanation of biological phenomena in terms of physics and chemistry. He freely acknowledged that the sciences of the inert world were relevant to an adequate understanding of life, but he denied that they were sufficient, hence the need for a biology which would explore highly complex organic phenomena. Lamarck's ideas about the existence of distinct degrees of complexity in nature have been mentioned several times, and it is a theme to which we shall return.

The word 'biology' was coined, almost simultaneously, by several writers around 1800. Lamarck's first published reference to it was in 1801, in *Hydrogeology*:

> A sound Physics of the Earth should include all the primary considerations of the earth's atmosphere, of the characteristics and continual changes of the earth's external crust, and finally of the origin and development of living organisms. These considerations naturally divide the physics of the earth into three essential parts,

Life

the first being a theory of the atmosphere, or Meteorology, the second a theory of the earth's external crust, or Hydrogeology, and the third a theory of living organisms, or Biology. (H 18)

The science of life only made sense in the context of parallel studies of the earth and the atmosphere. Living phenomena did not stand alone, but had to be seen as part of a larger whole, nature; indeed, they were only comprehensible when their constant interaction with the non-living world was recognised. The separation of living from inert nature enabled Lamarck to analyse more clearly their dialectical relationship. It was this sense of the relationships (Lamarck used the term *rapports* and continually emphasised links and connections) and interaction between various parts of nature which characterised Lamarck's mature work, and which led to many of his most original insights. He conceived of and carried out his project for the new science of biology in a spirit of synthesis.

Lamarck's understanding of how to approach life scientifically was governed by his reactions against other intellectual traditions. He came to oppose the commonly held idea of a vital principle, which he had used himself as a young man. It smacked of religion and of a restriction, the limitation of science's scope by declaring certain areas inappropriate for rational analysis. But equally, Lamarck condemned those who attributed to matter all the properties observed in the organic world. Matter itself had none of these properties; on this Lamarck was resolute. Life, then, was neither a being nor a thing, nor a species of matter. 'Life is . . . a physical fact. . . . Life is an order and state of things in the parts of every body which possesses it; it allows them to execute organic movements . . .' (R 70–1).

Into the deceptively simple phrase 'an order of things', Lamarck packed his philosophy of the organism. For him

Lamarck

there was only one kind of matter in the world, and, under certain conditions, it displayed organic properties. There was nothing whatsoever in matter itself which generated life, for matter was passive while life was active, according to Lamarck. The explanation of living phenomena lay in the relationship between matter and its surrounding conditions, and between the various elements within an organism. The naturalist should turn to laws of nature, and in particular to 'universal attraction which constantly works to bring particles of matter together, to form bodies, and to prevent their molecules from dispersing'; also, to 'the repulsing action of subtle fluids in a state of expansion; an action which, without ever disappearing, varies constantly in each place, and at every moment, and which changes in a number of ways the state of closeness of molecules in bodies' (A i 169).

The question of spontaneous generation was central to Lamarck's argument about the physical basis of life. He came to believe that, given the right external conditions and appropriately gelatinous matter, very simple organisms could be formed directly. Life originating in its most rudimentary form demonstrated how fine the transition was from a mass of jelly to a simple organism. Humidity, heat and gelatinous substance were all he thought it took for spontaneous generation to occur, not once, but continually. When organisms were so formed the impetus for life came from the environment. As living things became more complex, they depended less on environmental stimuli, because their organisational sophistication enabled them to display elaborate organic properties—the source of energy for vital action. These internal organic properties were also physical and amenable to scientific analysis.

Spontaneous generation had important implications for the history of the living world. His belief that the boundary between life and inert matter could be crossed did not touch

Life

Lamarck's conviction that there was a real hiatus between the two realms. In only two instances were transformations between the two domains possible—the spontaneous generation of extremely simple organisms, and the decay of living beings after death to form all the inorganic matter in the world. (Lamarck was, of course, aware that this could not explain the existence of matter in the first place. Life began, he thought, as a result of the spontaneous generation of simple organisms from pre-existing matter. Subsequently, matter constantly passed from an inorganic state to an organic one, and back again, in a perpetual cycle of nutrition and decay.) Living and inert phenomena were fundamentally different, yet the same matter circulated between the two realms—an aspect of Lamarck's belief in the unity of nature which will be discussed in the following chapter. Spontaneous generation demonstrated that the living world had a beginning which could be rationally reconstructed. This was an important step in the development of a transformist thesis which postulated a finite, temporal succession of organic forms.

The statement that life was a physical phenomenon, wholly within the aegis of laws of nature, constituted a methodological pledge on Lamarck's part. The organic world was special and remarkable, but it could be comprehended in scientific terms which aided rather than undermined the elucidation of the uniqueness of life. For example, whereas inert matter displayed only *properties*, organisms possessed *faculties*; 'faculties' being the name Lamarck gave to powers which he saw as deriving from the force we call life. The terminology illustrates the capacity for activity which Lamarck attributed to organisms. The essence of the organism lay in its ability to do things; ordinary matter just existed.

Animals and plants displayed two kinds of faculties. The first were general to all living things and required no

Lamarck

differentiated organs or organ systems. The simplest organisms were scarcely more than living blobs, 'animated molecules', yet they performed all the basic organic functions of nutrition and assimilation, the formation of their own bodies, growth and development, regeneration and reproduction. The second set of faculties, shown by a restricted number of organisms, were produced by specific organs: for example, wings produce flight. Lamarck divided this second category of faculties into two: 'constant' and 'alterable'. 'Constant' faculties were those of functional importance to the animal or plant concerned; hence they were not likely to be lost whatever environmental circumstances the organism was subjected to. They included digestion, for example by an alimentary canal, respiration by a special organ, feeling and intelligence, and sexual reproduction, which of necessity required specialised parts to produce eggs and sperm. (We may note that, although this analysis applied to both animals and plants, the examples Lamarck gave clearly had animals in mind, probably because they, more than plants, were active in all senses of the word, and so more 'vital'.)

By contrast, 'alterable' faculties were of less functional importance, and their development owed as much to environmental change as to the power of life itself. Alterable faculties changed with the organism's surroundings; they included locomotion by specific organs, touch, organs for holding and tearing prey, and for communication and defence. Constant organic faculties were produced by internal organs, while alterable ones derived from external, often appended parts. The former related to the life of the organism, the latter to the modifying influence of external circumstances. In terms of overall biological significance, alterable were subordinate to constant faculties.

The physical basis of life was further demonstrated by the role

Life

played in organic processes by so-called 'subtle fluids', the rarefied, all-pervading substances used to account for the phenomena of heat, magnetism and electricity. Eighteenth-century natural philosophy was based on the existence of such substances which were too fine to be either seen or weighed. Some of Newton's followers used them as the postulated causes of the fundamental forces in the universe, including gravity. Lamarck was not alone in treating such fluids and forces as virtually identical. Both could be known only by their effects. Subtle fluids exemplify eighteenth-century attempts to construct models and analogies connecting large and small-scale phenomena. Lamarck drew upon mainstream physics and chemistry in his use of subtle fluids which was inspired partly by his commitment to simple, economical explanations, and partly by his desire to make the inner recesses of nature easier to visualise and so more accessible to scientific understanding.

Lamarck used subtle fluids frequently when writing about physiological processes, as can be illustrated by his notion of 'orgasm'. He defined 'vital orgasm' as 'a specific tension in all the soft parts of living things, which holds their molecules a certain distance apart from each other, and which can be lost as a simple result of attraction between molecules when the maintaining cause of the tension ceases to act' (R 79). This state of tension enabled living things to display 'organic movement', one of the prerequisites for life. In animals, orgasm was the primary cause of irritability, their distinguishing feature. The orgasm could be of varying intensity, which explained how it gave rise to the contraction and distension characteristic of irritability. Although orgasm was present in the soft parts of animals, it was the fluids, both inside and outside the organism, which were crucial for its maintenance. Blood, or other fluids in invertebrates, upheld the tension of the orgasm, which was produced by the very fine, invisible, yet

Lamarck

potent fluids of the atmosphere which were responsible for maintaining life. This was yet another aspect of the interactions between organisms and their environment which was so fundamental a part of Lamarck's biology. In the simplest living forms these fluids penetrated the surface of the organism directly. In more elaborate life forms, the subtle fluids came from their food. In fact, Lamarck believed there was only one such fluid—caloric, the material basis of heat, of which electrical and nervous fluids were modifications—which made for a pleasing economy in explanatory terms.

Caloric was an important substance, according to Lamarck, who argued that virtually all physical and chemical phenomena could be explained by means of its properties or of those of its various states. His ideas about life were bound up with the chemical speculations with which he was deeply involved in the 1780s and 1790s. Lamarck was critical of the 'new chemistry', associated particularly with Lavoisier (1743–94) and Fourcroy (1755–1809), with its emphasis on elements, on quantitative analysis and on oxygen to explain the process of combustion. Theoretically, he allied himself to earlier chemical theory, which still enjoyed widespread support in the scientific community, where imponderable fluids in general, and caloric in particular, were fundamental.

Implicit in Lamarck's concept of orgasm was a model of organic processes as produced by the interaction of the solid and fluid parts of the body, where the former were passively acted upon by the latter. He explained nervous action in the same way by invoking nervous fluid, a form of electricity, which was conducted along the soft nerves. All nervous phenomena could be accounted for in this way—an explanatory device of some importance when it came to giving a convincing physiological account of how the human brain and mind worked.

Lamarck analysed irritability—the basic animal characteristic—in a comparative framework which combined his study of life and his theory of transformism. He was interested not just in life in the abstract, but in tracing the basic physical properties of living things from their simplest to their most complex forms. Among animals, life was expressed through the production of increasingly complicated forms, using irritability as the starting-point. More elaborate organisms developed over time on the template of earlier, simpler ones. The phenomena of orgasm and irritability manifested themselves in different ways according to the level of organisational complexity involved. In mammals, for example, irritability of the flesh lasted for only a few hours after death, while in the more primitive frog it might persist for twenty-four hours. Simple animals displayed irritability which was greater both in extent and in duration than that of their more sophisticated counterparts. The same pattern was followed by some other faculties, such as regeneration and reproduction.

Primitive animals were less internally differentiated than higher animals. In the latter, the organs became more specialised and restricted in their actions, and the organ systems more separate from each other as the number of functions the organism performed increased. Complex bodies were less capable of regenerating damaged tissues and of displaying great procreative powers. But at the same time they were better protected than rudimentary organisms against alterations in their milieu. It was, Lamarck thought, as if nature were compensating for the greater vulnerability of primitive animals to environmental fluctuations, by giving the substance of their bodies more resilience.

Since it was not possible to observe exceedingly fine fluids, other methods had to be employed. According to Lamarck, analogies and 'induction' were to fill the gap left by the limits of perception. To illustrate the importance of fluids in organic development, Lamarck developed an analogy between spontaneous generation and 'fécondation', by which he meant something close to our concept of fertilisation. The process of giving life through reproduction was like that of giving life to formless matter. In both cases, fluids were the agents of vitality—fecondating vapours and heat (caloric) respectively. The gelatinous substance out of which a new organism would grow contained no trace of life before the act of fertilisation. Lamarck therefore rejected the notion, prevalent in the eighteenth century, that embryos were preformed and required only a small stimulus to grow larger. Furthermore, he argued that matter had to be in a state of readiness to receive life, just like an egg before fertilisation. At that stage an egg was not alive, but merely prepared, in a purely physical sense, to make the passage to the organic state if, external conditions being right, fertilisation occurred. Reproduction and generation could be seen in a comparative context. In simple animals there was an interval between fertilisation and the first signs of life, while in viviparous animals, so Lamarck believed, the embryo displayed vital movements immediately. The movement of fluids played a crucial role in embryonic development; they were the agents of the changes which took place in the initially undifferentiated matter, for example, by separating parts which became different tissues and organs.

The supreme power of nature thus created living beings out of the passive, inert world. Lamarck insisted on both the spectacular powers of nature, and their physical character, a point exemplified by heat. Heat was intrinsic to life, it was the 'material soul of living beings' (R 102). To illustrate the intimate links between the two phenomena, Lamarck cited the earlier female puberty in hot climates, the capacity of heat to speed up life processes in general, and his belief that spontaneous generation occurred more

Life

readily in warm environments (R 99–100). Like life, heat had to be subjected to rigorous analysis to reveal its physical basis.

It is difficult to conceptualise so abstract a relationship as that between organisms and their environment. One way in which Lamarck did this was through the idea of atmospheric fluids. This was particularly appropriate for simple animals and plants where the environment was the major stimulating cause of organic processes. Lamarck did not see the environment as a hostile force, indeed, he attacked what he took to be the prevailing view that everything which surrounds living things tends to destroy them (R 72). The cause of deterioration was *within* each organism; it was the changing balance between assimilation from nutrition and loss by excretion. In an attempt to provide a physical model of the process of ageing, Lamarck postulated that the equilibrium between assimilation and excretion altered during the life of an organism. Until the peak of adulthood more was absorbed than lost; thereafter the opposite was the case.

The emphasis on the fine balance between two forces was thoroughly characteristic of Lamarck's natural philosophy. He employed the same approach in discussing the general relationship between organisms and their environment, that is, between living and inert nature. These two elements were construed as acting in opposition to each other, but Lamarck did not mean by this that the environment was 'hostile' to living things, nor did he ignore its role as the main stimulus of life in primitive organisms. Taken as a whole, nature was a harmonious system; none the less its constituent parts manifested different, contrary forces. In fact, there existed a dialectical relationship between organism and environment to which we shall return in chapter 6.

Lamarck

In examining Lamarck's definition of life and the implications of his assertion that it was a 'physical fact', we have inevitably touched on a number of the concepts which were constitutive of his transformism. The kernel of his biological philosophy is formulated in the following laws which clearly demonstrate how closely linked were Lamarck's transformism and his theory of life:

> First Law: By virtue of life's own powers there is a constant tendency for the volume of all organic bodies to increase and for the dimensions of their parts to extend up to a limit determined by life itself.
>
> Second Law: The production of new organs in animals results from newly experienced needs which persist, and from new movements which the needs give rise to and maintain.
>
> Third Law: The development of organs and their faculties bears a constant relationship to the use of the organs in question.
>
> Fourth Law: Everything which has been acquired ... or changed in the organisation of an individual during its lifetime is preserved in the reproductive process and is transmitted to the next generation by those who experienced the alterations. (A i 181)

Of particular importance was the third law, often referred to as the law of use and disuse, which encapsulates Lamarck's theory of the role of habit in organic transformations. It stated that parts of the body which are habitually employed will always develop, while those which are not invariably atrophy; and was mentioned by Lamarck as early as 1802 in his *Researches on the Organisation of Living Bodies*. The effect of habit on organisms had been widely acknowledged in the eighteenth century, and nineteenth-century evolutionists continued to regard it as integral to

Life

any understanding of organic change. According to Lamarck, organs grew and developed with use because the flow of fluids (blood, nervous fluid etc.) to the part increased; conversely, they atrophied without such use. Animals built up habits in this way through their active adaptive responses to their biological needs. Lamarck never claimed that animals exercise conscious choice or desire in building up habits, although many people have attributed such a view to him. The term 'need' has given much trouble to Lamarck's commentators (see p. 102): by it he meant quite simply the impulses to survive, the immediate reactions to stimuli, especially as they related to biological necessities like food, drink, procreation and the avoidance of death. Needs became more elaborate as animals became more complex. 'Habit' (in French, *habitude*) was a technical term in Lamarck's vocabulary; it provided a physiological explanation of adaptation to prevailing circumstances which, while demonstrating the impact of the environment on animals, also showed their active response to it. It followed that habits played a crucial role in the development of the animal series for, as the fourth law stated, modifications acquired as a result of them were passed on through reproduction, and persisted as long as there was adaptive value in them. Habit was a biological mechanism whereby simple organs became more complex and new organs developed.

> It is not the organs, that is the nature and form of the parts of an animal, which gave rise to its habits and specific faculties; on the contrary, its habits, its way of life and the circumstances in which the individuals from which it has descended, found themselves, have with time, constituted the form of its body and the number and state of its organs, and ultimately its faculties. (R 50)

Lamarck

The physical effect of habit was explained, of course, in terms of the action of fluids.

> Fluids moving in the soft parts of organisms which contain them, characteristically clear for themselves passages, settling places and outlets; they create canals and hence various organs; they vary the canals and organs either by different movements or by different fluids ... they enlarge, lengthen, divide and build up these canals and organs by the materials which form and which are constantly separated out from the moving fluids. (R 9)

It was the examples of the effect of habit which Lamarck gave that have been remembered: water birds developing webbed feet, the long legs and necks of flamingos and the neck of the giraffe. All these developments, he believed, stemmed from adaptations which aided survival and were then passed on because they continued to do so.

One of Lamarck's most striking examples of the powerful effect of habit was the difference between the helpless newborn child and the adult at the peak of development. He believed that organs had to be used if they were to grow properly. This example also highlights an ambiguous feature of Lamarck's argument. He slid from referring to changes in individuals over short periods of time, to changes in species over longer periods, to the whole animal series developing over aeons of time. There might have been a creative use of analogy here whereby different processes were illuminated through comparison. Alternatively, Lamarck may have simply been making the assumption that individual adaptations added up to the transformations of species over thousands of years. Lamarck's arguments imply that he did not come to a firm conclusion about the scale of organic change, preferring to leave the questions raised by such analogies open.

Life

The new science of life which Lamarck envisaged embodied the premiss that organic and inert phenomena were fundamentally different. At the same time he continually drew upon his earlier ideas about physical and chemical processes to explain how organisms functioned. Furthermore the environment played a central role in the generation and maintenance of life. None the less, once alive, organisms displayed faculties never observed in the world of brute matter. Special laws and axioms therefore had to be invoked as the theoretical foundations for the new science of biology.

Lamarck's notions of life and of biology cannot be understood without accepting the historical importance of 'analysis' and the methodological orientation which Lamarck built upon it. Starting with complex animals, he traced the origins of their organ systems in simpler forms. Then, beginning with rudimentary creatures, he showed that the levels of organic complexity were also historical stages in the production of the animal kingdom. It was the task of biology, the science of life, to study these levels.

Biology purported to deal with those features plants and animals shared. Lamarck never wrote his planned treatise on biology but produced instead the *Zoological Philosophy* (1809). There he discussed the nature of life at length; when he compared plants and animals, it was the latter category which gripped his imagination and led him to stress zoology rather than biology. He continued to be interested in plants, believing that transformist ideas should be rigorously applied to them, and using botanical illustrations in his arguments. But in reality, Lamarck's biology did not serve to link the study of plants firmly to that of animals; instead, it put zoology on a broader footing. His zoological philosophy connected natural history, including classification, physiology and the medical sciences, to the sciences of the environment.

6 The sciences of the environment

Interest in the environment as an object of study increased markedly during the eighteenth century. There were a number of reasons for this and they shed light on Lamarck's interest in the milieu of living things. As a result of geographical exploration, the variety of natural objects was becoming better known in this period. Much of the diversity among plants and animals in different parts of the globe was understood in terms of the effect of climate on organisms. The impact of such external conditions on living things was amply demonstrated by the problems caused by emigration and moving plants and animals from one environment to another. Such transitions were clearly perceived by medical practitioners, whom the high reputation of Hippocrates, an early exponent of environmental medicine, had stimulated to study the impact of the physical world on human beings.

It was an integral part of Lamarck's utilitarian purpose in studying science that the worst effects of the environment would be avoided if the precise factors which affected organisms were successfully identified. The potential for control and improvement was generally agreed to be considerable, especially where man and his illnesses were concerned. Medical investigation was immeasurably aided by the attention given in natural philosophy to factors thought to be responsible for organic change, such as the fine and exceedingly subtle fluids Lamarck made so much of.

Environmentalism was also strongly reinforced by those philosophers who rejected innate ideas in favour of a model of the human mind in which all knowledge came from the

The sciences of the environment

senses. Through this route the environment made an impact on the life of each person. We know that Lamarck, as a follower of Locke and Condillac, was acutely aware of the biological implications of their views. Just as experience acquired by the senses moulded human understanding, so organisms were moulded by their milieu. Lamarck explored the reactions of organisms to their surroundings, utilising his notion that organic beings were plastic and responsive. This, indeed, was a view common in eighteenth-century physiology in the wake of Albrecht von Haller's work on irritability and sensibility, the biological properties which came to symbolise reactivity (see p. 29).

Lamarck's interest in matters connected with the environment in general and with its impact on life in particular, was therefore not unusual. What was noteworthy was his ability to weld scientific analysis of the environment to an understanding of the organic world within a historical framework. His very definition of biology had implied the existence of two additional, parallel sciences devoted to the earth and the skies respectively. In addition to geology and meteorology, astronomy and mineralogy may be grouped together as sciences of the environment, that is, those disciplines which study the non-living parts of nature. The role of these fields in Lamarck's work was to explain how the characteristics of living things were produced through interaction with their surroundings. The way Lamarck construed the science of biology demanded complementary studies of the environment; conversely, studying the physical world without taking account of its dynamic relationship with living things seemed to him pointless.

Lamarck's early enthusiasm for meteorology introduced him to a wide range of natural philosophical subjects, including heat, light and colour, electricity and magnetism, and it encouraged him to think about such problems as the nature of gravity, laws of attraction and repulsion of matter,

Lamarck

the moon and the tides. Studying the weather appealed to that contemplative side of him which found natural history attractive. It required detailed observation of a wide range of phenomena which were to be arranged taxonomically in just the same way, and using the same methods, as plants and animals. What a natural history of the weather really entailed was revealing the regularities beneath the apparent randomness of nature. Classification also helped to further some of the most important goals of meteorology: the improvement of agriculture and medicine (many diseases were thought to be environmentally caused) and giving aid to travellers, soldiers or sailors, where their safety depended on weather conditions.

Meteorology was a field where untrained observers could play a valuable role in taking regular and careful readings of the most important scientific variables. Lamarck hoped that all their results would be pooled by a central state-supported agency and tried unsuccessfully for many years to find help for such a venture. The full measure of his Englightenment optimism manifested itself here. He was not in the normal way an optimistic man, and the poor reception his meteorological writings received caused him much pain, but something of his idealism persisted. It came out clearly in his conviction that in meteorology co-operative research, done mostly by amateurs (in the true sense of the word), was both desirable and possible, and was the best way to advance the science.

The use that Lamarck made of the environment in his mature work went far beyond a long-standing interest in the physical world. His notion of life, the centrepiece of his philosophy, was based on a dynamic relationship between responsive organisms and a potent milieu. Since he believed that organic forms were profoundly affected by accidental events in their physical surroundings, no science of biology would be complete if it failed to take account of the

The sciences of the environment

intricate interactions between organisms and their environment. This was the source of his interest in the environment, and it explains the importance he attached to the collection of empirical material on the earth and atmosphere.

Lamarck approached the environment in a number of different ways. He examined the effect of physical agents in the atmosphere, such as electricity and other imponderable fluids, using chemical theories as his main tools. Temperature, humidity and winds also had to be charted. In addition he considered geographical factors, like rivers, mountains and seas. By taking account of changes over long periods of time the slow transformations which the earth's crust had undergone could be analysed.

The impact of the environment on life was therefore seen in terms both of its short-term effects on individuals, and of its long-term effects on the history of living nature. We have already noted Lamarck's tendency to conflate individual life history with the history of species (see p. 56). Although there may have been a deliberate use of analogy, whereby the former was taken to be a mechanism for the latter, Lamarck often wrote as if there were no differences between the two levels. Certainly his thoroughgoing conception of a unified system of nature held together these different ways of approaching the environment. Along with a view of nature united by a set of universal laws, Lamarck held a firm belief in economy of explanation. Since the same forces operated in different realms of nature, it was simply bad science to invoke different explanatory terms if the same would serve.

The unity of nature was also apparent in other aspects of Lamarck's environmentalism. Minerals, he argued, were produced by decaying living matter, making a pleasing cyclical relationship between living and inert nature. 'Without exception, the raw compounds which form most of the earth's external crust and continuously modify it by

Lamarck

their changes all result from the remains and residues of living organisms' (H 91). When an organism died, there was nothing to hold the complex substances together any more and they gradually disintegrated, only to be recombined by new living beings, in a perpetual cycle. It was, in fact, one of the distinctive things about bodies which possessed life that they could form combinations out of simple substances. Physical compounds disintegrated into their constituents when left to themselves, and only living things, particularly plants, had the power to bring elements together to form combinations. 'The organic action of living organisms creates combinations of substances which would never otherwise have existed' (H 85).

Minerals could be understood in terms of the historical sequences they passed through; they were not, Lamarck insisted, permanently fixed species, as naturalists asserted, but temporary states. Assigning names to every different substance observed was a fruitless exercise. Of course, Lamarck could not explain the origin of life in these terms. The first organisms produced by spontaneous generation were made of pre-existing matter. Lamarck stressed that his cyclical view was not intended as an explanation of origins but as a description of currently operating processes (see p. 47).

As suggested earlier, Lamarck's interest in origins served a very specific, limited purpose. It was useful as part of his strategy to demystify man, life and nature, by showing that they had finite beginnings which could be comprehended. But, unlike his mentor Buffon, Lamarck was not particularly interested in the origin of the earth. Rather, he wanted to demonstrate that the earth was in a state of perpetual change. 'On our planet, all objects are subject to continual and inevitable changes which arise from the essential order of things' (H 61). Where the earth was concerned, changes did not add up to produce a specific goal or act progressively,

The sciences of the environment

as they did for living things. Lamarck wanted to convince his readers that all matter was in flux and, equally importantly, that change took place very, very slowly. He followed the remark cited above by saying:

> These changes take place at a variable rate according to the nature, condition, or situation of the objects involved, but are nevertheless accomplished within a certain period of time.
>
> Time is insignificant and never a difficulty for Nature. It is always at her disposal and represents an unlimited power with which she accomplishes her greatest and smallest tasks. (H 61)

These statements implied that the earth was much older than many dared think. 'Oh, how very ancient the earth! And how ridiculously small the ideas of those who consider the earth's age to be 6,000 odd years' (H 75).

It is clear how important vast periods of time were for Lamarck when we consider his bold thesis that the seas have moved at least once all the way round the globe. Every point on the earth's surface had been covered by water at some time in its history—a hypothesis capable of explaining the presence of fossils in remote inland areas, such as the remains of marine forms on mountains. In this way he avoided resorting to universal deluges to explain the distribution of fossils, and provided an alternative explanation to universal catastrophes and extinction. Lamarck bolstered his arguments against extinction by assuming that the world was very ancient, and that there had therefore been plenty of time for transformations to have taken place. This was another instance of the influence of Buffon, who in his *Theory of the Earth* (*Théorie de la Terre*, 1749) had dramatically increased the age of the earth. Lamarck's rejection of catastrophes and extinction, and his belief in the antiquity of the earth were facets of his larger

Lamarck

commitment to a transformist account of living nature in which organisms changed and adapted slowly over vast stretches of time.

Lamarck placed great emphasis on fluids as fundamental and potent natural forces, so it was consistent that he should see water as the major agent of geological change, as the title of his book *Hydrogeology* implied. Nor is it unexpected that the language used to describe the action of water on the globe was the same as that to depict the actions of fluids in living bodies: erosion, carving out channels and sedimentation. The emphasis on water and other fluids seems odd only when the central place of fire in Lamarck's chemical theories is recalled. He showed no awareness of heat as a principal agent of geological change, though it did play a part in his discussions on the position of the equator. He postulated that the equator moved along with the seas, so that the same geographical area had experienced a variety of climatic conditions over the aeons of time the earth has been in existence. Alterations in climate nevertheless took place gradually, allowing living species to adapt and avoid extinction by migration or by transformations.

Since earth history was not moving in any particular direction, the impact of the environment on living things could not be such as to produce regular, graduated series of animals and plants. Instead, the environment led to organic diversity, deviations from the regular progression of organisms produced by the 'power of life'. Thus the milieu might accidentally produce specific adaptations in organisms, but it did not contribute to the increase in their complexity which Lamarck attributed solely to the 'power of life'. It followed that the history of nature was open-ended, the outcome impossible to predict, except in so far as organisms continued to become more complex. 'Man is the latest result and present climax of this development, the ultimate

The sciences of the environment

limit of which, if it is ever reached, cannot be known' (H 77). Lamarck thought of organisms and their environment as the products of two opposite forces, analogous to attraction and repulsion. Both the physical and living worlds had histories. Their histories were even intertwined, yet their patterns were different. Organisms changed so as to become more complex, while there was no direction in environmental change.

Lamarck's early enthusiasm for studying the weather had an appreciable impact on the direction of his thought. Similarly, fierce debates on extinction and catastrophes prompted him in the 1790s to think out his approach to fossils, to change in both inorganic and organic nature, and to the role of time as an explanation of environmental change. In many ways issues raised by the study of the environment were an important stimulus for Lamarck's transformism. *Hydrogeology* (1802) and the *Meteorological Annuals* (Annuaires Météorologiques) (1799–1810) contained important views on the nature of life which reveal something of Lamarck's development in this crucial period when he was consolidating his transformist position.

Meteorology attracted Lamarck for a number of reasons, some of which have been mentioned earlier. He saw it as a test case for the formation of a new scientific discipline, and attempted to outline its methodological priorities: another example of the integral importance of questions of scientific method in Lamarck's natural philosophy. Furthermore, Lamarck identified similar key issues in botanical and meteorological classification. Meteorological facts were to be recorded and then arranged in classes, orders, sections and genera according to the principles of natural classification. A precise pattern of seasonal variations would then emerge. The beauty of the approach was that 'here there are no systems, no hypotheses, no danger of error: the facts

Lamarck

alone, properly classified, determine everything' (M vi 13).

Lamarck divided the causes of atmospheric conditions into two types: those which are variable and irregular in their action, and those which follow 'regular and progressive laws'. The second related principally to astronomy, especially the moon and sun which acted on the atmosphere through universal attraction. Lamarck described meteorological events in terms of a monthly cycle made up of two fourteen-day units. The phases of the moon were extremely important in his view, but so also were geographical factors—mountains, coasts, rivers, valleys and forests—which affected the 'climatic constitution' of a given area. Thus, the structure of Lamarck's explanation of meteorological change was identical to his explanation of organic change. In both cases regular laws of nature operated, but their actions were veiled by random, accidental conditions which were unpredictable.

Prediction of weather conditions was not possible, Lamarck recognised, because of their production by these two kinds of causes. But he believed that the regular laws which operated entitled him to assess the *probable* weather for a given latitude at a specific time of year. Furthermore, if the regular laws had not yet been fully understood, it was simply a matter of there being insufficient accurate observations. The method for establishing the new discipline of meteorology was clear—it consisted of organising an army of observers taking climatic data according to a uniform, rational plan, and then subjecting the information to statistical treatment. From these procedures, Lamarck was confident, would emerge the theories and general principles of meteorology. He set out the priorities for the new science. Its adequacy depended, in decreasing order of importance, on instruments, the precision of readings, the exact timing of readings, and a good method for dealing with the facts gathered. There was also a scale of importance

The sciences of the environment

for meteorological parameters, with air pressure, wind direction, and temperature being the most important, followed by electricity, humidity, water evaporation, rainfall and magnetism.

Two features of Lamarck's meteorology prejudiced its reception. First, the emphasis on the moon, and the use of terms such as 'the zodiac', were strongly reminiscent of astrology. The same impression may have been given by the second factor, his emphasis on forecasting. Lamarck complained in the third meteorological annual (1801) that his talk of probabilities had been misconstrued as claiming the power of prediction. An anecdote about the reception of the annuals is instructive. It is said that Lamarck tried to present a copy of a book to Napoleon, who spurned it publicly, on the grounds that it was just another of those despised almanacs. The volume was in fact the *Zoological Philosophy*, his masterpiece!

When examined from the perspective of his biology, Lamarck's meteorology fitted into a recognisable pattern. Recommending his annuals to agriculturalists and medical practitioners, he pointed to 'the influence of the atmosphere on vegetation' as 'an undoubted fact and widely recognised', and claimed that 'the power of the atmosphere on the animal economy is not to be denied'. Environmental changes had quite specific physiological effects of which Lamarck gave several examples:

> variations in the weight of the atmosphere, which impinges upon us from all sides, increase or diminish the tone of our organs, and hence speed up or slow down the circulation of body fluids ... fluctuations of temperature open or close the routes by which transpiration takes place ... changes in atmospheric humidity either rob us of our natural heat or maintain it ... (M ii 12–3)

The full range of Lamarck's meteorology covered plants,

Lamarck

animals and human beings. He advocated that scientific societies, especially those concerned with agriculture, should record whether germination, growth, foliage and flowers and the maturation of fruits and seeds were advanced, retarded or impaired; similarly the quality of crops, the behaviour of spiders and migratory birds, and information about epidemics should be recorded. The project for a science of meteorology was thus closely linked with the project for a science of life.

Lamarck's environmental sciences were built on his early chemical ideas. We have already seen that he rejected the approaches to chemistry associated with Lavoisier, especially what Lamarck saw as a tendency to invent more 'species' of substances, and a complex nomenclature to match. He reasserted a simple Aristotelian scheme, with four elements—earth, air, fire and water. All compounds were, according to him, constantly in a state of change moving towards their simple constituents. By contrast, the theories which he opposed and called 'pneumatic' (air-based) as against his own 'pyrotic' (heat-based) ones, postulated that matter in its simplest form had a strong tendency to form combinations with other substances. Lamarck explained many chemical reactions in terms of the different physical states of fire, the material cause of heat.

According to Lamarck, there were so many possible permutations of matter which were always temporary, that it seemed to him fruitless to establish a rigid nomenclature; the arbitrary assigning of names was something to which he was always opposed. There were thus a number of important similarities in his approach to inert substances on the one hand, and plants and animals on the other, in his use of classification and in his assumptions about mutability over time. Yet the differences should also be noted. Although biological species were, like chemical compounds, temporary,

The sciences of the environment

they lasted for long enough and represented sufficiently natural groupings (as reproductive units) to be worth naming. Inert matter was for Lamarck inherently more volatile than organisms. He was critical of the 'new chemistry' for its experimental methods which, by subjecting substances to violent tests, changed them into *'unnatural'* compounds which generated little information of scientific value. Chemical experiments therefore exacerbated the problem by artificially generating substances.

Chemistry was fundamental to an understanding of all the physical alterations of both inert and living matter which Lamarck wished to catalogue. The argument of *Hydrogeology* relied heavily on his theories of mineral formation from plants and animal bodies. The detailed discussion of such questions as colour, the nature of blood and other body fluids, combustion and fire which figured prominently in the books Lamarck published in the 1790s, assumed less importance from 1800 onwards when his main interests were in the transformations the organic world had undergone. Some of the underlying structures of thought remained, such as the ubiquitous action of fine, invisible, weightless fluids, as important explanatory devices. The key theme, nature in flux, endured.

Once again we have found that questions of scientific method lay at the heart of Lamarck's thought. For example, the fact that nature was constantly changing posed problems, since this was hard to observe directly. He was convinced of the essential similarities between past and present processes, and enunciated, as a methodological principle, that the two must be analysed in the same terms. This led him to reject global catastrophes (which had never been observed), but not local ones, which had; he rejected a universal deluge, not small-scale flooding, and admitted that volcanic action might have limited effects. Lamarck defended the creative use of hypotheses in areas like earth history where direct

Lamarck

evidence was so scarce. He claimed that since complete knowledge was impossible in several areas of science, the goal to aim for was a *plausible* account which contradicted no known facts, and which tried to give a coherent explanation of processes which would otherwise remain mysterious.

Straightforward observational work gave Lamarck much pleasure. His efforts to classify and name clouds, to distinguish different layers in the atmosphere, and to assess the effects of diurnal climatic changes, were of considerable scientific value. In geology, Lamarck's direct observational knowledge was limited. He had had little field experience, since his only period of travel had taken place before his interest in geology was well developed. He was most familiar with the Paris basin, through his work on the fossil invertebrates of the region.

In developing his knowledge of the physical circumstances of life, Lamarck used mineralogy, chemistry, geology and meteorology. What emerged was a sense that the same explanatory principles were to be employed in all of these sciences. Simplicity and economy of explanation were a priority. A greatly expanded timescale, continuity between past and present, and the conviction that nature was never immutable were the foundations of transformism—for which the insights derived from the sciences of the environment were prerequisites. His concept of the historicity of nature was the basis of Lamarck's most significant intellectual achievements.

7 Transformism and the *Zoological Philosophy*

Lamarck arrived at his theory of the transformation of organic forms in 1799–1800 in the context of heated debates on extinction and fossils. In 1800, the opening lecture of his course at the Muséum revealed his new-found belief in the mutability of living nature. The undoubted novelty of his ideas and the controversies they provoked should not distract us from the equally important point that the roots of transformism went back to his earliest scientific work.

His theory rested on the following propositions: nothing in nature is constant; organic forms develop gradually from each other, and were not created all at once in their present form; all the natural sciences must recognise that nature has a history; and the laws governing living things have produced increasingly complex forms over immense periods of time.

Lamarck drew on numerous instances of transformations in both the inorganic and organic realms: the tides, chemical and geological mutations, spontaneous generation, processes of learning and development, ageing and adaptation. By 1800, the belief that 'nothing in nature is immutable' was a basic axiom in his natural philosophy. Despite their distinctive organic characteristics, the historical changes plants and animals underwent were only one aspect of the flux of nature.

The most complete and best-known formulation of Lamarck's transformism was the *Zoological Philosophy* of 1809, which placed transformism in a broad biological

Lamarck

context. The work must be seen as an ensemble, and an appreciation of its overall structure illuminates Lamarck's ideas of organic change. The *Zoological Philosophy* brought together classification, an analysis of the nature of life, especially in simple animals, and an account of the complex behavioural capacities of higher animals. Little evidence remained of the palaeontological or geological preoccupations of seven years before. Apparently Lamarck no longer felt the need to rehearse the arguments against extinction and in favour of continuous environmental change. His mind was, by 1809, firmly fixed on two projects: a natural history of invertebrates, and a study of man with particular reference to the nervous system and ethics. This twin focus, on the simplest and the most complex parts of the animal kingdom—central to transformism—was everywhere manifest in the *Zoological Philosophy*.

It was divided into three parts, each dealing with a distinct aspect of Lamarck's theories of living things: the natural history of animals, the physical causes of life, and the physical causes of what he called *sentiment*. Although this is best translated as 'feeling' or 'sentience', it is crucial to recognise that it was the biological capacity to receive sensations which concerned Lamarck, and not conscious acts.

The first part reassessed the classification of animals. From his botanical work, Lamarck was aware of the problem of imposing a system of classification on the natural world and then treating it as if it derived from nature itself. While other naturalists might have been content merely to acknowledge the artificiality of their systems, Lamarck was not. He strove to develop a way of coherently ordering animals while following nature's own plan. The project took on new significance with his transformism which offered a strong underpinning for a more natural classification because of its capacity to

Transformism and the Zoological Philosophy

determine the actual order in which organisms were produced. For Lamarck, classificatory schemes ought to express the real relationships between living objects. Transformism and taxonomy were not mutually exclusive; although species were mutable they should still be named and their relationships with other forms specified. Before classifying, it helped if one had a theory to account for both the differences and the similarities between animals. Lamarck attributed the differences largely to the accidental effect of environmental factors. The similarities derived from 'the power of life', a law of nature which produced higher animals out of lower ones. Nature's use of basic templates to generate organic series of increasing complexity explained the observed relationships between forms.

Before Lamarck, it had been customary to begin classificatory schemes with the most complex animals, gradually descending towards the more simple ones, as, indeed, he himself had done in early writings. Following the pattern he had established in the *System of Invertebrate Animals* (1800) and the *Researches on the Organisation of Living Bodies* (1802), Lamarck devoted a lengthy chapter (P i 130–217) to the 'degradation and simplification in organisation from one end to the other of the animal chain, going from the most complex towards the most simple'. The concept of degradation was not a new one and would have been familiar to many of Lamarck's readers from Buffon's *Natural History*. Lamarck used it to suggest that, taken as a whole, the animal series displayed a striking gradation between complex and simple, from those with many faculties, a skeleton and vertebral column, to those entirely lacking these features. Having established the idea of a linkage between the main groups of animals, the 'principal masses' (Lamarck was careful to say that he was not concerned here with *species*), he reversed the direction of the series, making a chain of decreasing complexity into

Lamarck

one of increasing complexity, starting with the most simple animals. The *real* history of the animal kingdom was conveyed by this new sequence. Natural history was now truly the history of nature.

Lamarck's ultimate goal was to understand the plan nature had followed and thereby to discover uniform, constant natural laws. He implied in early sections of the *Zoological Philosophy* that it was a law of nature to produce ever more complex living things which displayed regular, fine relationships between them. He was aware, of course, that the animal kingdom was not like that, and that the discrepancy was increasingly apparent the more one examined families and genera, rather than classes and 'principal masses'. Whereas nature's laws were responsible for the gradations among living things, the action of the environment accounted for specialised adaptations. Lamarck had given a naturalistic account of the relationship between organisms and their surroundings in *Hydrogeology* (1802), where the language he used had suggested an opposition between living nature and the environment, so that the latter produced 'anomalies' in animals (P i 134–5). As we have seen, for Lamarck the environment was a major part of nature; it operated according to natural laws, yet it was also in some sense the antithesis of life. 'Life' was the special power of nature to produce ever more elaborate, integrated and active organic beings. The inorganic matter which made up the physical environment, left to itself, would decompose into its simplest constituents (see p. 62). Hence the natural world was composed of two forces constantly interacting in a dialectical manner; for an accurate classificatory system to be arrived at, they had to be unscrambled.

The influence of the environment was more evident in some cases than in others. Families of genera and species were groups with only fine gradations of organisation between members, there having been no extreme environ-

Transformism and the Zoological Philosophy

mental changes to cause greater differences. Indeed, this seemed to act as a definition of 'family' as a taxonomic grouping. Had it acted without impediment, the 'power of life' would have produced a succession of regularly graduated forms starting with simple ones. Hence, if one looked at the general series of animals as a whole, the impact of the environment was clear in any deviation from this pattern.

Having argued in the first part of the book that species were not fixed, that animals could be arranged on a scale of increasing complexity which used human beings as the standard, and that classification should follow the chronological order of nature and hence be as 'natural' as possible, Lamarck set the scene for the analysis of the most important concept of the book, that of life.

The purpose of the second part of the book was to show that life was a purely physical phenomenon and to sketch out some of its basic properties. It therefore set out the first principles of biology with particular reference to zoology. The analysis of life was of considerable explanatory importance, for transformism rested on a number of presuppositions about the properties of the organic world, and 'the power of life' was itself a mechanism of transformation. Lamarck outlined his technique of finding out 'what life really is' by looking at simple animals with no special organs.

Between excited and communicated motion Lamarck drew a distinction the importance of which cannot be exaggerated: vital motion acted by excitation, motion in inert bodies was by communication. The definitions expressed his belief that although life could be analysed in physical terms, different physical principles should be invoked to explain inert matter. When motion was transmitted from one physical object to another, it was quite permissible to speak in terms of cause and effect for these could in fact be clearly separated. This was not the case in the living world, where cause and effect were inextricably

Lamarck

intertwined as the nature and speed of the operations of the vertebrate nervous system illustrated. It should be emphasised that Lamarck did not thereby abdicate his responsibility as a scientist to find a physical explanation; he was merely asserting that the inorganic and organic worlds worked in different ways. Lamarck's sense that cause and effect could not be neatly separated in the organic realm was consistent with his emphasis on the dialectical relationship between organism and environment, and between different parts of the organism itself.

Locating the *source* of vital stimulation was Lamarck's next step. He thought that in simple animals it was the imponderable, invisible fluids in the environment, while in the most perfect animals, the excitatory cause of life was within each individual. Lamarck, following eighteenth-century medical and physiological traditions, located it principally in the nervous system. The environment as a source of vitality was exemplified by the spontaneous generation of rudimentary organisms. Lamarck stressed the role of fluids as mediators between organism and environment, as agents of all vital actions and as the mechanism whereby the number of organs and their associated functions increased. The superior vital energy of higher animals was manifested in their fluids, especially in nervous fluid, which acted on their passive parts (what Lamarck called 'containing' parts), the cellular tissue. All these remarks laid the foundations for the hierarchical distinctions Lamarck wished to make, for example between animals and plants and between vertebrates and invertebrates.

It will be remembered that Lamarck took irritability as his starting-point in the search for the distinctiveness of the animal kingdom. He suggested that nature progressively added additional faculties on to irritability: first, the capacity to have sensations, then a primitive consciousness, and finally, the ability to perform complex mental functions.

Transformism and the Zoological Philosophy

A classificatory scheme of the whole animal realm was implicit even when Lamarck discussed physiological generalities. Thus, at the end of the first part of the *Zoological Philosophy* Lamarck provided an overview of the animal kingdom arranged in six 'degrees of organisation', each reflecting a different level of sophistication of responsiveness—a hint of the preoccupation with the nervous system and its faculties found in the third part of the treatise.

Lamarck was struck by the capacity of higher animals to change and adapt their behaviour through a highly complex nervous system. In human beings, it was the extraordinary capabilities of the brain and nervous system which characterised the species. Man—the masterpiece of nature—served as a vivid illustration of how the most intricate vital processes function. Lamarck therefore devoted the third part of the *Zoological Philosophy* to 'the physical causes of sentience', a section close to three hundred pages long in which, starting with first principles, he set out his approach to the analysis of nervous phenomena, including the operations of the human mind. His most important premisses have already been mentioned: the rejection of innate ideas, the belief that all experience comes from the senses, and the assertion that structure and function are indissolubly linked. This part of Lamarck's masterpiece sometimes embarrassed subsequent generations because its emphasis on mental phenomena appeared to give weight to the commonly held view that Lamarck had illegitimately attributed consciousness and will to all animals, and hence had been guilty of psychologising biological phenomena.

It was in the *Zoological Philosophy* that the influence of an important contemporary intellectual movement can be most clearly discerned, that associated with the *idéologues*, as Napoleon had pejoratively dubbed them. Following Condillac, and closely identified with the philosopher Helvétius (1715–71), the mathematician Condorcet (1743–94)

Lamarck

and their families, the members of this loose network devoted themselves to developing a naturalistic analysis of the human mind—a science of man. Their position was threatened by Napoleon's closure in 1803 of that part of the National Institute (the successor of the Academy of Sciences) devoted to the moral and political sciences. Their ideas were liberal and anti-authoritarian, with a strong streak of rationalism; very close, in other words, to the political views we believe Lamarck held. He also shared their commitment to developing a science of man, and saw in their project an important role for the biological sciences, that of demonstrating the physiological basis of mental events. The leading exponent of ideology in the biomedical sciences was P. J. G. Cabanis (1757–1803), whom Lamarck quoted with approval as an authority who affirmed the importance of organisation (i.e. anatomy and physiology) as the unique source of both moral and physical attributes. By 'moral', Lamarck and Cabanis meant a high degree of organic complexity, as in mental phenomena, but no value judgement was implied. While Cabanis confined his work to man, Lamarck extended the *idéologue* perspective to the whole animal kingdom, in order to demonstrate that the apparently special abilities of the human race were based on the gradual, historical development of organisms.

Lamarck's ambitious project to build a science of man on the basis of a science of zoology required a notion of levels of organic complexity, which we have previously discussed. He criticised the *idéologues* for their exclusive emphasis on the most complex form of life—man—and their neglect of nature's more modest products. What had been for them a largely philosophical exercise he wished to make into a biological one:

> In seeking evidence for the fact that the moral and the physical have a common source, one should not confine

Transformism and the Zoological Philosophy

oneself to the study of the highly complex structure of man and the most perfect animals. One will find it far more easily through a consideration of the various progressive increases in organisational complexity from the most imperfect to the most complicated animals. This progression reveals the origin of each animal faculty, and the causes and manner of development of these faculties ... One will then be convinced yet again that the two major aspects of our existence, called the physical and the moral, which present two such apparently distinct orders of phenomena, have their shared source in organisation. (P i 364–5)

Lamarck's earliest work in botany had embodied the idea that identifying degrees of complexity was an essential step in developing an adequate classificatory system. This was achieved by establishing which were the most and the least complex forms, and then by filling in the space between them through assessing whole plants, not just isolated parts. He subsequently applied the method to zoology: 'if the lower end of this scale displays the minimum of animality, the other end necessarily displays the maximum' (R 38). The levels of complexity in plants were less striking to Lamarck than in animals; plant activity and life was relatively impoverished. The most vivid example of structural levels of complexity was provided by the animal nervous system. Lamarck did not say that one nervous system developed directly out of a previous one, but he pointed out that anatomical parts were added on in more complex animals, giving rise to more elaborate capacities. Anatomical, physiological and taxonomic levels were identical.

Some of the components of Lamarck's transformism were general applications to biology of approaches which the *idéologues* had applied only to man. For this he required a comparative perspective. Part three of the *Zoological*

Lamarck

Philosophy achieved a synthesis between a number of different intellectual traditions, and it also showed an ability to assimilate new styles of thought. Lamarck did not stagnate in middle age, and the lively dialogue he entered into with the *idéologues* in his writings proves the point.

When it came to mechanisms for transformism, Lamarck quite unselfconsciously assumed the inheritance of acquired characters, in the sense in which he understood it, to be so obvious and unexceptionable as to require very little comment. Since antiquity it had been believed that adaptations to changes which had taken place during the lifetime of an individual would be passed on to their offspring. Scientists continued to employ the idea, and the closely related one of habit, long after Lamarck's death. In addition to the inheritance of acquired characters, he simply postulated that nature was constantly in change and that life, by its very nature, became more complex with time. The environment certainly played a role as an agent of change. These ideas had been set out in his *Researches on the Organisation of Living Bodies* in 1802 and they remained the foundation of his later writings. It should be noted that for Lamarck transformism was a logical consequence of his views on the nature of the organic world; it was arrived at by deduction from the fundamental axioms governing all living things rather than by induction from a large number of empirical examples. This is not to say, of course, that Lamarck did not have a deep fund of natural historical knowledge on which to draw, for clearly he did. It was rather that since he regarded transformism as proven, he did not feel obliged to go and search for instances of it.

Adaptation was an empirical phenomenon which was important for Lamarck's arguments, and he understood it to be built into the nature of organisms. Biological need, grounded in the interaction of life and environmental

Transformism and the Zoological Philosophy

forces, was a stimulus for action in animals without any 'effort' or 'will' being involved. The straightforward absence of the organ systems necessary for consciousness meant that most animals reacted to prevailing conditions by instinct rather than intelligence—a faculty which, according to Lamarck, nature had distributed with exceptional parsimony. The drive to adapt was so strong that animals responded automatically to stimuli from the outside world, and from inside their own bodies, like thirst or hunger. The emphasis on adaptation gave a teleological cast to Lamarck's arguments in that much stress was laid on the purposiveness of living things. His repeated personification further heightened the sense of nature having goals. Not only did it seem as if there was a pre-ordained purpose in nature, but Lamarck's language also implied that nature worked, even laboured, in the service of specific ends. This impression was an unfortunate product of Lamarck's rhetoric. In fact he thought there was no purpose outside nature, and human beings were merely one species among the many nature had produced. What commentators have construed as teleology, Lamarck saw as adaptation and progress generated by the interaction of physical forces.

The real achievement of the *Zoological Philosophy* was its fusion of a number of hitherto distinct areas—natural history, classification, physiology and psychology. It was the direct product of Lamarck's project for a treatise on biology, in that life and its unique characteristics were at its centre. At the same time, the mutability of organic forms was simply one example of uniform, natural laws which governed change in all bodies. The distinction between life and non-life notwithstanding, Lamarck saw nature as a single system of natural laws.

The *Zoological Philosophy* was rich in biological ideas which never minimised the complexity of living things, for Lamarck well knew that wholes were more than the sum of

Lamarck

their parts; hence his emphasis on organs and organ systems as they actually functioned. He expounded a number of biological axioms of great importance not only for his own scientific work but for both contemporaries and successors who were involved with general and comparative physiology and with theoretical biology. The most significant of these were the tight link between structure and function, the interacting powers of life and the environment, the law of use and disuse, and the inheritance of acquired characters. The discussions of the *Zoological Philosophy* were impressively broad in scope, for they encompassed the nature of matter and of life, the role of fluids in living and inert bodies, and the distribution and classification of animals. Lamarck continually stressed that nature took time in her productions and was unable to do things at a stroke; he predicated transformism on a system whereby complexity accreted stepwise on a simple base. These themes were illustrated on the one hand by simple invertebrates and on the other by the complex nervous systems of vertebrates. Lamarck extended his study of lower animals in his seven-volume *Natural History of Invertebrates* (1815–22), which he regarded as furnishing the supporting evidence for the zoological principles he had enunciated in 1809. Complex living forms were also dealt with in his *Analytical System of Man's Positive Knowledge* of 1820, an attempt to move from first principles to a naturalistic ethics with the human race and its future as the principal focus.

8 Nature and God

Although in the eighteenth century the idea of nature was of undisputed importance, there was no consensus on what was meant by the term. Nature could be seen as active or passive, as a fount of moral values or as violent and cruel, as the glorious product of God's creative power or as a system of natural laws owing little or nothing to the deity. Lamarck's whole conception of science was based on his beliefs about nature. For example, his search for natural methods of classification implied that nature contained coherent patterns which scientific activity could reveal. The belief that the principal patterns embodied linked series and graduated chains may be found in most of his work, and it played a central role in the development of his transformism. Lamarck sketched out his ideas about the essential characteristics of nature in the *Zoological Philosophy*, and developed them considerably in the lengthy theoretical introduction to the *Natural History of Invertebrates*. He made his philosophy of nature most explicit in an article on 'Nature' for a prominent natural history dictionary (D xxii 363–99), which was reprinted in its entirety in the *Analytical System* (s 20–96).

In placing particular emphasis on his definition of 'nature', Lamarck followed eighteenth-century traditions of natural philosophy. General concepts like 'nature', 'natural law' and 'God' were important for two reasons: the first was metaphysical and was concerned with what existed in the universe, at both physical and spiritual levels; the second was methodological and was concerned with what knowledge it was possible for the human mind to acquire. The philosophy of nature was therefore closely related to the

Lamarck

practice of science. Science was knowledge of nature gained by the human mind. The characteristics of both nature and the mind were thus germane to the tasks of science. It should be remembered that for Lamarck human understanding was itself a product of nature and natural laws, and not in any way separate from them.

Transformism gave him the tools to deal with these two concerns in a coherent way, because it combined an account of empirical reality with a theory of knowledge. The theory that species have undergone transformations during the history of the world supposed that nature worked in gradual steps, building diversity on simple templates. One could only know this by looking backwards from the complex to the simple, finding the units of action and structure appropriate to each level of complexity, and recomposing them in a natural taxonomy which recreated, using the human mind, what nature did with matter. The belief that nature worked that way also provided Lamarck with an explanation of mental processes—that knowledge is gained by means of sensations, linked together to form ideas. The psychological capacities of human beings were grounded in the physiological faculties of the nervous system, itself a product of transformism.

There was a further reason why the definition of 'nature' was urgent for Lamarck. He did not see nature as passive, as many earlier natural philosophers had; on the contrary, in his biological work in particular, he was continually seeking to explain the activity organisms displayed. Since he eschewed soul, spirit, vital principles, and a constantly interfering God as explanations, he had to look elsewhere. He had to look, in fact, at the properties of nature itself.

Lamarck's definition of nature was deceptively simple:

> Nature is an order of things made up of objects extraneous to matter. These can be known through the

observation of the physical world. As a totality, nature constitutes a power which is unalterable in its essence, determined in all its acts, and constantly acting on all parts of the physical world. (D xxii 377)

While he freely acknowledged God's unlimited creative power, Lamarck did not see this as having any bearing on natural science. By contrast, the natural laws which produced physical phenomena were the proper explanations natural philosophers should employ.

Nature was not the totality of physical matter—Lamarck called that the universe—nor was it the final cause, for that was God. Nature was a system of laws, which together with motion, and using space and time, produced all the bodies human beings perceived around them. The concept of 'production' is significant. Lamarck distinguished production from creation, which was an instantaneous act of supreme power; the prerogative of the deity and which nature could not aspire to. In emphasising nature as a productive power, Lamarck suggested that much time, and even labour, was involved in the history of nature. In this respect transformism was at one with his notion of nature. Nature was an abstract general term which stood for all active forces, and it was sharply differentiated from passive matter.

These active forces were part of a system; nature operated in consistent ways and was a harmonious, balanced and stable totality. Nature's harmony is well illustrated by Lamarck's emphasis on biological adaptation, not on conflict and competition. The question of harmony took on added importance when Lamarck applied his ideas to human society—a noticeably unharmonious phenomenon. He searched for ways in which society could display the harmony of nature, seeing consensus and co-operation as the proper goals of civilisation. The balance of the natural

Lamarck

world was visible in the range of species which peacefully coexisted and in the equilibrium between inert matter and organic forms. Organisms were finely attuned to prevailing environmental conditions. Significantly, the balance was only disrupted, according to Lamarck, by human beings, as when they drove a species to extinction.

How are we to reconcile Lamarck's reference to nature's 'stability' with his repeated assertions that nothing is constant in nature? Flux characterised the constituent parts of nature, but nature as an ensemble gained an overall stability from the permanence and deterministic character of natural laws. Laws had necessary effects, and there was no possibility of their changing except at the will of the Creator, which meant, for Lamarck, that they would not change at all.

In Lamarck's words, nature was 'blind', by which he meant that we cannot attribute to it intentions or conscious goals. Will, in this sense, was the exclusive right of God. He held the Creator totally separate from nature; to do otherwise, he explained, was to confuse the watch with the watchmaker. Lamarck was only interested in the watch, and although its very existence implied that of a maker, one could get perfectly good service out of a watch, and study it in detail, without giving its maker a moment's thought. He invoked a Creator to explain the origin of nature and the universe, but this was no active, interfering or constantly present God. This was a God of convenience who filled an explanatory gap—the origin of the world—which would otherwise have been left empty. For the purposes of science, therefore, God was to be disregarded, and Lamarck viewed with deep suspicion those who did otherwise. There is no evidence of deep religious belief on Lamarck's part, nor of a crisis of conscience in coming to his transformist beliefs, for he was, at heart, a secular rationalist whose most profound commitment was to the study of nature. Science

Nature and God

must take the mystery out of nature, not least because such mystery could all too easily be used for repressive social and political purposes were it to remain unchallenged. Unfortunately, Lamarck did not specify how this use might occur. This was one of the reasons for his rejection of universal souls and vital principles: the obscurity such concepts introduced could all too easily be misused. Lamarck's ideal was for science to be an instrument of mental improvement, a tool of reason not of religion.

His rationalism was taken a step further. When man saw chance or disorder in nature, he was revealing his own ignorance, which led him to apply such labels to things he was annoyed by or could not understand. Chance and chaos had no part in lawful nature. Although acknowledging the study of nature and the human condition to be inextricably linked, Lamarck believed that people should try as best they could to avoid projecting their prejudices on to nature.

The drive to comprehend nature was of the greatest importance, he considered, for the future of the civilised world, precisely because human beings were part of nature. Natural history was the highest imaginable calling, not just because nature was full of wonderful things, the contemplation of which was uplifting, but because it was the only route to self-knowledge. Enlightenment thought placed much emphasis on self-knowledge; hence attempts to promote a science of man, a project deeply sympathetic to Lamarck.

Lamarck further insisted that people have no real creativity or imagination; all ideas come from nature, with the consequence that transcendence of the material world is impossible. Man is of nature in all senses, and he must use his position to improve the lot of the human race as a whole through the scientific understanding which, alone among nature's products, he can acquire.

It would be a mistake to assign a label to Lamarck's

Lamarck

metaphysical position, although many scholars have been tempted to do so. It is clear, for example, that he was no crude materialist. Nor, except as a young man, was he a vitalist—one who explains biological phenomena in terms of 'vital principles'. He did not worship nature—although he did call her 'the communal mother'. The key to his natural philosophy was his concept of laws of nature acting on matter to produce complex, structurally elaborate forms. The characterisation of his approach in terms of naturalism is thus quite apt.

Lamarck drew from his definition of nature clear-cut conclusions about the manner of acquiring knowledge. There were observable regularities, produced by natural laws, which it was the business to scientists to record and classify despite the difficulties of observing infinitely slow changes. Observation was, however, the only route to secure knowledge, no matter how difficult it might be. The sense of the unity of nature which Lamarck expressed was connected with other aspects of his methodological credo; simplicity and economy of explanation, and the interconnectedness of different parts of nature. These were in fact two faces of the same coin; the same limited number of explanatory terms should be employed because nature always acted in the same ways, and her various productions were linked because they developed gradually, one after another, over a period of time. It followed for Lamarck that science should not study isolated objects but seek out the relationships, analogies and affinities between the constituents of a larger ensemble: nature.

9 Man

Human beings presented a particular challenge to Lamarck, since he had pledged himself to explain all physical objects as products of nature. In his time it was common to view humanity as bearing the special stamp of God's handiwork. Lamarck, like the *idéologues*, was committed to developing a secular account, a science of man, as an alternative to such views. Focusing on the being who was uniquely both a product of nature and capable of comprehending it by scientific means, he completed his plan for a unified natural philosophy with a discussion of how the human mind worked. Because of his long-standing interest in scientific method, which inevitably raised questions about mental processes, Lamarck's search for the nature of the human being might be said to go back to his earliest work in natural history and to classification in particular. Understanding human thought was an essential prerequisite for being a good scientist, for only then would the inherent limits of knowledge be revealed, allowing man to set his sights accordingly. From the relatively modest and conventional attitude of *Researches into the Causes of the Principal Physical Facts*, (*Recherches sur les causes des principaux faits physiques*) published in 1794, but written in the late 1770s, where he stated that the vital principle could never be understood, Lamarck moved towards a position of greater confidence, bolstered by his new conviction that no such things as vital principles existed. He believed that everything except God was amenable to scientific analysis.

In *Researches on the Organisation of Living Bodies* (1802), Lamarck's first sustained discussion of the nature of

Lamarck

man, the gap between human and animal was acknowledged. But any reassurance which the orthodox might draw from the statement was immediately undermined by the assertion that there was a gradation of human intelligence which was acquired and not inborn. Those who failed to develop their mental capacities, the ignorant, were little better than animals—and they were in the majority. Lamarck refused to draw an absolute distinction between man and animal any more than he did between mind and body when he located their common base in organisation, i.e. physical structure.

Lamarck went on to note the amazing similarity between the orang-utan and man. He made his position even clearer when he devoted a lengthy section of the book to the nervous fluid as the explanation of will, thought and other mental phenomena. Admittedly, when it came to classification, man had his own order, genus and species; yet his uniqueness was not absolute but merely relative to other animals, and, as such, had a physical basis. 'This special state of human organisation has been acquired little by little as a result of much time and with the help of favourable circumstances' (R 134). There can be little doubt of his intentions: they were to repudiate totally an account of mankind inspired by religion, and to substitute a transformist perspective. This entailed rejecting mind–body dualism in favour of understanding living nature as the product of historical processes.

Like other organisms, human beings were moulded through interaction with their environment by the effect of habit on their bodies. Lamarck liked the proverb 'habits form a second nature' ('les habitudes forment une seconde nature'—P i 237), and suggested that the less intelligent a person was, the more rigid his acquired habits would be. Intelligence itself was a product of the habit of thought; for Lamarck found the human brain to be the organ most

Man

responsive to constant use and development. The mechanism by which such physiological changes were effected was the nervous fluid. The special properties of this animalised form of electric fluid enabled Lamarck to provide an organic basis for nervous processes without having recourse to models of transmission, such as vibration, which might be appropriate for inert matter, but not, in Lamarck's opinion, for living things. Vibration, a fashionable model in the seventeenth century, was based on an analogy with inanimate objects, like the strings of musical instruments, and was altogether too mechanical a conception for Lamarck. He sought to develop and apply a biological language to the most complex mental acts of which human beings were capable.

The nervous fluid participated in two movements, carrying impulses from the central nervous system to the muscles, and bringing sensations from the periphery (the sense organs) to the centre (the spinal cord and brain). Two additional functions took place in the most developed nervous systems. The first was denoted by the untranslatable term *sentiment intérieur*—literally, inner feeling or sensation, but carrying no connotations of consciousness and including a range of functions less elaborate than conscious thought. Thought was the second additional function of which the most elaborate nervous systems were capable, and it was the final capacity added to the nervous system in the course of the history of nature. Over the span of the animal kingdom, these four functions, the two movements of the nervous fluid, *sentiment intérieur* and thought, were formed successively. Since no function was possible without the appropriate structural substratum, the progression was reflected in a series of anatomical parts, starting with medullary masses, then a spinal cord and simple brain, up to the large cerebral hemispheres of human beings, where all higher mental actions took place. While

Lamarck

the brain was the seat of intelligence and thought, the spinal cord interacted with muscles and maintained basic organic functions without the participation of the brain. The fine details of anatomy held no interest for Lamarck, whose project was the development of plausible models and analogies for processes which were, by definition, impossible to observe directly.

Lamarck's approach to the nervous system assumed a chronological sequence, in which levels of organisational complexity had been added progressively. The levels of processes and relationships, being part of nature, were embodied in structures composed over time. It went without saying that such levels also reflected classificatory groupings.

In analysing the nervous system, Lamarck extended these levels to internal organic functions. For example, he postulated a number of distinct centres which co-ordinated different kinds of nervous action to replace the traditional notion of a *sensorium commune* (common centre)—the seat of the non-material soul. Rejecting free will as an illusion and presenting reason and will as merely products of judgement—one of the basic faculties of the understanding—Lamarck employed no idea of a soul or of a single seat of action. He used the idea of functional levels to show that, even in human beings, most responses did *not* demand conscious thought, but were instinctual. Thought and instinct coexisted in harmony because of the functional complexity of the nervous system; it ensured the production of integrated behaviour necessary for adaptation and survival.

When he applied his biological ideas to ethics and politics, Lamarck argued that the social problems he saw around him had their origin in human nature itself. In an analysis strongly reminiscent of Rousseau's famous argument about inequality, he suggested that man had become distant from

Man

nature, his proper home. Man's capacity for individuality served him badly, for it led to the differences between people which engendered inequality, and that in turn entailed the domination of one social group by another. Lamarck explained the potential destructiveness of human nature in terms of a number of 'propensities' which were part of man's psychological make-up. He subjected these propensities to an elaborate taxonomic analysis, as he did all mental acts. Following his similar dissection of the plant, animal and mineral kingdoms, he broke down complicated, seemingly inscrutable wholes into simple, more manageable parts.

In the first volume of his great work on invertebrates, Lamarck included a section in the introduction 'on propensities, of both sentient animals, and of man himself, considered with respect to their origins, and as phenomena of organisation' (A i 259–303). He unfolded his analysis of the propensities of social man in the context of the whole animal kingdom and as examples of animal faculties. Propensities were the causes of passions, in turn the source of human destructiveness. The main human propensity was that towards self-preservation, followed by self-love, a tendency to seek well-being, a tendency to dominate and a repugnance towards one's own destruction. Each of these was made up of subcategories, so that under self-love, he distinguished *amour propre* from egoism. He put forward ten moral maxims to counteract destructive propensities and passions where, once again, the theme of restoring nature's equilibrium and order came to the fore.

We may briefly note the political consequences of these ideas. The keynote of Lamarck's social philosophy was the need for harmony and balance among people, and between human society and nature. This would be achieved, he believed, by sticking closely to natural laws, reconciling the interests of different sections of the community, and by

Lamarck

achieving a balance between public and private interests. Rule by a small, disinterested meritocracy, with natural scientists in pride of place, was Lamarck's ideal. He had little faith in the majority of his fellow citizens because he saw them as caught up and deluded by the artificiality and duplicity of society. Their ignorance of nature told against them. While Lamarck yearned for a society more in tune with nature, he certainly did not envisage rejecting the knowledge society had acquired. On the contrary, he was critical of Rousseau's negative attitude towards science, preferring to see science as the instrument which would liberate social man from oppression and degradation and so hasten his return to the laws of nature.

There was a positive, if limited role for religion to play in society. For reasons entirely explicable by natural laws, the human race was unique in its capacity to experience pain and suffering and to fear death. Ideas of God and immortality comforted a species so exquisitely conscious of its own existence, and, for this reason alone, should be tolerated. Similarly, the exercise of the imagination should also be tolerated, except in relation to science. Lamarck found notions of absolute truth meaningless, and truth was certainly not guaranteed by God, as Descartes had argued. The idea of 'God' was, in fact, the product of human intelligence, an invention. God was no longer a necessary truth or a product of sheer faith. He saw a way round the argument that God must exist, since human beings possess the idea of Him: for Lamarck, human beings arrived at ideas such as infinity, omnipotence and eternity by first thinking of their *opposites*, each of which were a part of everyday experience. He further argued that concepts of right and wrong were relative, not absolute, and changed according to their setting. These statements reveal his methodological stance; all the ideas we take to be absolute have to be analysed into the experiences from which they are composed.

Man

Nature was also appealed to as an arbiter of moral matters. Yet even this most fundamental tenet was not without its difficulties when applied to humanity. Lamarck saw a paradox: men and women, unlike other organisms, did not always follow natural laws; hence his injunction about the need to get back to a more natural state. But natural laws were all-powerful and all-determining; escape from them was impossible. The human race was free, or rather, freer than other animals, and simultaneously subject to nature. As we have seen, this situation led to political problems and social man had to be deliberately led back to the nature it was impossible to evade. A similar complexity was raised by the existence of suicide. Since Lamarck postulated as one of the basic human propensities a universal aversion to death, the explanation of suicide posed a problem. 'Suicide is the result of an unhealthy state in which the ordinary laws of nature are inverted' (s 226). But how could the laws of nature change? Lamarck gave a physical explanation of suicide as caused by a disruption of the nervous system, through fever for example. Thus he saw suicide as a form of illness, and exonerated victims from responsibility for their actions. Lamarck was clearly torn between his desire to provide physical explanations for all observed events and his belief that acts such as suicide were unnatural. For him, as for others committed to naturalism, there was no simple solution to the paradox of human contravention of incontrovertible natural laws.

At an individual level there were clearly moral lessons to be learnt from Lamarck's natural philosophy. He stressed the necessity for education and mental activity to build up good habits of thought. An overactive mind was as undesirable as a sluggish one, since nervous fluid could easily become depleted with excessive cerebration. Denoting the special make-up of an individual by the classical term 'temperament', Lamarck stressed that this, like the physical

Lamarck

products of habits in general, could be inherited. In effect, he applied transformism to all aspects of human existence. Nowhere was this better illustrated than in his account of the faculties of the understanding—attention, thought, memory and judgement. For each one he described its underlying physiological processes. The most interesting, from the point of view of Lamarck's biological thought, was attention, because it illustrates his continued insistence on the activity of organisms. Attention prepared the organ of thought for action: without it, sensations could not be properly registered as thoughts. The organism had to fix its attention in an appropriate way for environmental stimuli to be received; its stance towards its surroundings was an active one, often prompted by internal biological needs.

> When you reflect or are preoccupied by something, although your eyes open and the objects around you continually impinge on your sight by the light they emit, you see none of these objects, or rather, you do not distinguish them. This is because the effort, which constitutes your *attention*, directs the available nervous fluid towards the traits of the ideas with which you are preoccupied, and because that part of the organ of intelligence which is such that it can receive the impressions of the sensations generated by these external objects, is not at that moment ready to receive these sensations. Thus the external objects which strike your senses from all sides give rise to no ideas ... But if your *sentiment intérieur*, stimulated by some need or interest, suddenly directs the nervous fluid to the spot in your organ of intelligence to which the sensation of the object in sight is related ... then your attention prepares that part of the brain to receive the sensation of the object which affects us, and you will acquire some kind of idea of this object. (P ii 393–5)

Man

This quotation demonstrates the commonsense flavour of Lamarck's writings. Daily experiences, like sleep, dreaming, fevers, and manual skills, were appealed to as sources of insight into human nature. He also considered that introspection was as reliable a method as any other: it was based on observation and, given the particular problems of studying the human condition, it was indispensable.

The emphasis on the nervous system was consistent with the writings on the science of man by the *idéologues* and their associates which Lamarck was reading just before he produced the *Zoological Philosophy*. He extended their arguments by stating that the science of man, which sought to locate both the physical and moral aspects of human existence in organisation, was the proper domain of the zoologist, who could place the human being in the temporal hierarchy of animal forms. Only if man were seen in the context of the whole animal kingdom was his special state comprehensible. Lamarck was by no means uncritical of the philosophical tradition which emanated from Condillac, and he differed from its exponents on many important points, most of which stemmed from his particular concern to apply their ideas to the comparative physiology of different groups of animals.

Lamarck's attempts to conceptualise mental and nervous processes were significant for a number of reasons. He employed concepts (for example, that of 'habit') which became a focus of debate on mind–body relationships in nineteenth-century French philosophy. He also successfully negotiated the treacherous path between a crude mechanistic reductionism and a traditional religious dualism because his style of thinking compellingly promulgated ideas of a dynamic, progressing natural world, in which complex mental phenomena emerged from biological development. Lamarck's work demonstrated a thorough-going

Lamarck

naturalism, the dominant creed in nineteenth-century scientific thought.

The idea of man was clearly at the centre of the *Analytical System of Man's Positive Knowledge* of 1820, which has been described as Lamarck's intellectual credo. The book was divided into two parts; the first dealt with objects which existed *outside* man, the second with organic systems *within* man. Little of the material and few of the arguments were new, although Lamarck drew out the ethical, social and political implications of his natural philosophy more explicitly than before. For the first time, taxonomy faded into the background and the human race in its unique position as part of nature and as the knower of nature became his main focus. The picture of human life and society was a sad, cynical and disillusioned one, where self-interest motivated all acts and where artificiality had come to predominate over nature. It is hard to resist a biographical explanation of Lamarck's response to humanity. By 1820 he was old, blind and poor, and felt unappreciated by the majority of his scientific colleagues. He believed his originality had not been recognised sufficiently, and thought that the vested interests of his rivals, notably Georges Cuvier, who wielded considerable political power, were to blame.

Lamarck brought to his understanding of man many of the conceptual tools he had developed over the previous twenty years or so. His appreciation of the human race as part of nature was firmly rooted in his theory of transformism and he consistently explained the development of society and the evolution of language in terms of increased needs leading to new forms of communication which in turn produced yet more new needs. Furthermore, he recognised the cultural variety of human civilisation as a historical product of interaction with a variety of environments, and did not assign to any single group the privilege of being

Man

more 'natural' than the others. Lamarck's transformism led him to cultural relativism.

Lamarck's approach to the human race reveals facets of his thought consistently maintained over a long, productive career. Even in his earliest taxonomic work he saw that the most perfect forms shed light on less perfect ones. In this sense, man was an integral part of his scientific method. The real goal of science for Lamarck was knowledge of nature, because it offered self-knowledge to humanity.

10 Lamarck's legacy

Lamarck's ideas were nothing if not controversial. They were invoked by countless nineteenth- and twentieth-century commentators of different nationalities and creeds, and played a particularly crucial role in the evolutionary debates of the second half of the nineteenth century. While a full study of the scientists and philosophers who have discussed Lamarck is beyond the scope of this book, some of the more important among them should be mentioned briefly. A discussion of these reactions to Lamarck may serve to put the inheritance of acquired characters into a more balanced perspective, since Lamarck has generally been known, and become notorious, on account of this idea alone.

From previous chapters it should be clear that the inheritance of acquired characters was just one proposition in a complex system of natural philosophy. It certainly played an important role in Lamarck's explanation of the transformations organic beings undergo, but it is seriously misleading to take it out of the context of his biological thought as a whole and out of the historical setting in which he developed it. During Lamarck's lifetime it was a commonplace that these habits, and the characters they gave rise to, which an individual acquired were passed on to its offspring. The idea only became contentious after the German embryologist August Weismann (1834–1914) in the 1880s had purported to offer empirical refutation of the idea. Lamarck and Lamarckism were thenceforth associated with a particular scientific controversy; many of the participants in it had never read a word Lamarck himself wrote. The fact that they constructed a mythical Lamarck

Lamarck's legacy

to suit their own concerns is the key to understanding Lamarck's legacy. It was the mythical Lamarcks—for there were several such images elaborated—that captured the imagination of post-Darwinian biologists: Lamarck the genuine, but neglected, founder of evolutionary theory; Lamarck the French Darwin; Lamarck the discoverer of purpose in the universe; Lamarck the materialist.

How images of Lamarck were constructed can be illustrated by the *éloge* given after his death by Georges Cuvier, the famous zoologist and palaeontologist who was Lamarck's colleague at the Muséum. It was customary for the perpetual secretary of the Academy of Sciences to give an account of the life and achievements of academicians when they died. In this case the 'eulogy' quickly achieved notoriety for its offensive and condescending treatment of a man who was without dispute a natural historian of great eminence. Far more explicitly critical than was usual on such occasions, it ridiculed much of Lamarck's work as mere speculation, and was careful to praise only his most detailed taxonomic work. So vitriolic was the 'eulogy' that it was read and printed only after Cuvier's own death in 1832.

Cuvier's attack on Lamarck hinged on two main points. First, he suggested that Lamarck's work was not firmly grounded on an empirical base but derived from fanciful speculation. In its drive towards synthesis it was, Cuvier claimed, divorced from real scientific progress, which was always based on increasing disciplinary specialisation. Along with Cuvier's thesis about the desirability of restricting the scope of scientific disciplines went his conviction that some questions, such as the nature of the human mind, were, *a priori*, outside the realm of science. Lamarck held the contrary view, that scientific investigations should not be bounded by arbitrary criteria; all natural phenomena were legitimate objects of enquiry.

Lamarck

Cuvier's second point of attack concerned the boundary between mind and body, a well-established area of fierce scientific, ideological and philosophical debate. Lamarck found such a boundary to be of no operational value in the biological sciences; hence he considered the distinction itself to be scientifically invalid. Cuvier accused him of misconstruing the mind–body relationship to such an extent that he attributed to animals a will and consciousness found only in human beings. Cuvier thus inaugurated the popular misconception that Lamarck explained evolution by the conscious striving and efforts of animals. It is one of the ironies of history that Lamarck had accused Cuvier of precisely the same mistake in claiming that sensibility was a general feature of the animal kingdom. For Lamarck, sensibility was a specific function, and therefore based on an equally specific organ or organ system. Clearly, his argument continued, such structures were not universally present in animals, many of whom experienced no sensations of any kind, for their organisation was too rudimentary. The source of the trouble was Lamarck's term *besoin*, which meant either 'want' (in the sense of desire) or 'need'. The difference between these alternatives is crucial. Lamarck intended the term in its second sense, to suggest the biological imperative or drive which led animals to adapt to changing environmental conditions to ensure their survival. His sense of the fine, gradual increments in the complexity of the nervous system over the animal kingdom as a whole made it unlikely that he would attribute consciousness to all its members.

Cuvier intended his 'eulogy' of Lamarck to be a warning to those who wished to indulge in scientific speculation on such questions as the mutability of species. It has been suggested that his real target was Étienne Geoffroy Saint-Hilaire, who was sympathetic to Lamarck's ideas, shared many of his views and believed evolutionary change could

Lamarck's legacy

occur as a result of direct and sudden environmental changes acting on foetal development. In 1830, the year after Lamarck's death, Cuvier was involved in a well-publicised debate with Geoffroy at the Academy of Sciences. Geoffroy's work concerned the detailed anatomical relationships between organic forms, especially their skeletons. He argued that the principle of unity of composition should be applied to comparative anatomy to show how all members of the animal kingdom were modifications of one basic anatomical type. While this did not logically entail an evolutionary theory, Geoffroy increasingly inclined to a theory of descent, paying particular attention to the environment as the agent of organic change. Predictably enough, Cuvier was thoroughly hostile to such views, which were tainted by association with German *Naturphilosophie* and hence to his mind as speculative as Lamarck's. The Cuvier–Geoffroy debate was important because it led to a belief in a natural alliance between Geoffroy's and Lamarck's ideas despite their differences, and because it affected the way in which *The Origin of Species* was received in France.

English scientists learnt of Lamarck largely from the detailed critique of his ideas provided by Darwin's mentor Charles Lyell in *Principles of Geology* (1830–3). Lyell's view was not unlike Cuvier's; that Lamarck simply did not begin to provide convincing empirical support for transformism. Lyell added that Lamarck's handling of fossil evidence was particularly weak, since he was unable to prove a historical progression of species. Lyell's arguments are of special interest, partly because of his close relationship with Darwin. Also, it is possible to compare what he said in print with his private notebooks, which make very clear what it was in Lamarck's writings that stuck in Lyell's throat. It was Lamarck's insistence that man be seen as an integral part of nature's history, and his assertion that there

Lamarck

was a real continuity between man and the animals. For Lyell, Lamarck thereby destroyed the clear boundary which sustained the separate orders of the natural and social worlds. One can sense Lyell's despair when he anticipated that there would be 'no line of demarcation between rational and irrational, responsible and irresponsible' if Lamarck's ideas were true.

The security of Lyell's categories depended on a strong definition of immutable species, on successive creations, and on the separation of man from the rest of nature. And yet, when it came to explaining environmental change, Lyell, like Lamarck, was a uniformitarian who used laws inferred from his own observations to explain the past and assumed no fundamental differences in the intensity or force of past and present processes. In fact, Lyell was perfectly correct in his assessment of the implications of Lamarck's thought; Lamarck's naturalism certainly aimed to break down traditionally held boundaries, and he had little reverence for his own species. Eventually, Lyell was converted to an evolutionary position, and felt he had done Lamarck an injustice, but by that time the reception of Lamarck's ideas was inextricably linked with the fate of Darwin's theory of evolution by natural selection.

Reactions to Darwin were profoundly affected by the anonymous publication in 1844 of *Vestiges of the Natural History of Creation*. In fact written by the autodidact and publisher Robert Chambers (1802–71), it presented an evolutionary view of nature which ranged from the development of the solar system to embryology. *Vestiges* provoked an outcry, yet it used contemporary scientific knowledge too skilfully to be simply dismissed out of hand. Chambers was certainly familiar with Lamarck's ideas, yet he was not uncritical of them. Although commonly associated with Lamarck, Chambers developed a very different theoretical structure which drew on detailed geological and embryo-

Lamarck's legacy

logical evidence—something Lamarck's work never did. Chambers was deeply influenced by natural theology; hence the overall thrust of his book bore little resemblance to Lamarck's writings. None the less, *Vestiges* drew evolutionary biology in its broadest sense to the attention of the reading public, made the topic a familiar one and incidentally made Darwin a good deal more cautious in the way he expressed himself.

After *The Origin of Species* was published in 1859, the *fact* of evolution became widely accepted. Subsequently it was the *mechanism* of evolution which principally concerned scientists. Darwin by no means settled the question of how evolutionary development occurred. He postulated as probable causes both natural selection and the inheritance of adaptations to environmental change.

Lamarck may be understood as the most significant forerunner of Charles Darwin (1809–82) in the sense that he stated a number of principles that were fundamental to the latter's work. Lamarck's belief that species were not immutable entities produced by special acts of Divine creation was based, as was Darwin's, on a scepticism derived from the difficulties of taxonomy. Both had found it extremely difficult in practice to make clear discriminations between closely related plant and animal forms. Both men used fossil evidence to argue for the long, complex history of life on earth. They were further united in their conviction that nature should be treated as a single system—science was to be an enterprise of synthesis. But if we go beyond these shared beliefs, we find profound differences between Lamarck and Darwin which were produced by their different world-views.

Where Lamarck postulated a tendency inherent in living things to become more complex as time went on, Darwin did not see random variations in life forms as necessarily leading to a progression. Lamarck denied that species

Lamarck

became extinct, since their plasticity enabled them to adapt to changing external conditions. Darwin, more conscious of the limits of organic variation, found extinction to be a common feature of nature's history. Indeed, the existence of traces of extinct forms confirmed Darwin's evolutionary beliefs. Lamarck believed in the spontaneous generation of simple organisms from inert matter, as did many nineteenth-century biologists; Darwin denied that this was possible. Where Darwin found animals and plants competing for scarce resources and struggling to survive, Lamarck perceived adaptation and adjustment, phenomena which convinced him of a fundamental harmony within the natural world.

Darwin himself disclaimed any direct connection between his ideas and those of Lamarck, while recognising a certain community of interest: 'Heaven forfend me', he wrote, 'from Lamarck's nonsense of a "tendency to progression", "adaptations from the slow willing of animals", etc! But the conclusions I am led to are not widely different from his.' While Darwin and some of his supporters emphasised their distance from Lamarck, the same could not be said of the man who was arguably the most influential evolutionary theorist of the century, Herbert Spencer (1820–1903). It was reading Lyell's critique of Lamarck that made him into an evolutionist:

> Why Lyell's arguments produced the opposite effect to that intended I cannot say. Probably it was that the discussion presented, more clearly than had been done previously, the conception of the natural genesis of organic forms. The question whether it was or was not true was more distinctly raised. My inclination to accept it as true ... was, doubtless chiefly due to its harmony with that general idea of the order of Nature towards which I had, throughout life, been growing. Supernaturalism, in whatever form, had never commended

Lamarck's legacy

itself ... Hence, when my attention was drawn to the question whether organic forms have been specially created or whether they have arisen by progressive modifications, physically caused and inherited, I adopted the last supposition.

The use Spencer made of Lamarck's ideas illustrates the manner in which evolution was part and parcel of social theory—a point of some importance, since Lamarck clearly saw the social realm as governed by natural laws, and his ideas were attractive to Spencer precisely because of the link between biological and socio-cultural evolution that they encouraged. Spencer's evolutionary world-view made a particularly deep mark upon American thinkers, which may perhaps account for the greater sympathy for Lamarck shown there.

Despite Spencer's early espousal of evolution and Lamarck, it was the explosion of controversies which followed the publication of *The Origin* which had a most dramatic effect on attitudes towards Lamarck. Far from relegating him to the august position of a remote intellectual ancestor, the controversies drew his ideas to the attention of many people who had never previously heard of him. Lamarck probably had most impact between 1860 and 1910, the period when arguments about the mechanism of evolution were at their fiercest. Those who commented on Lamarck generally focused on four issues of enduring biological importance: man's place in nature, holistic views of the organism, the nature of adaptation, and the role of the environment as an agent of organic change. The inheritance of acquired characters and the respective contributions of heredity and environment to an organism's final form were specific cases of these larger concerns.

Those who discussed Lamarck did so in a number of different ways. To some extent he became an accepted

Lamarck

alternative to Darwin, his name frequently being invoked to promote an aspect of biology which Darwin, or more often the Darwinians, were thought to have neglected. The term 'Lamarckian' can be applied to those who took up Lamarck's ideas, whether consciously or not. One should not, however, expect all Lamarckians to share the same beliefs. Since Lamarck was interpreted in a number of often contradictory ways, Lamarckism likewise took on many different aspects. Such terms are employed by historians to suggest a kinship between others and Lamarck, but they were also used polemically, to abuse or to praise, during the evolutionary debates themselves.

The designation neo-Lamarckian is more specific and is most properly restricted to a group of American naturalists who, beginning in the 1860s, explicitly associated themselves with Lamarck. The movement was strengthened in the 1880s through the need to stave off Weismann's attacks on the inheritance of acquired characters and by the neo-Darwinism of Wallace and Weismann which asserted the sufficiency of natural selection as the sole mechanism of evolution. Many neo-Lamarckians were reacting against the anti-evolutionism of Louis Agassiz (1807–73), the influential Harvard zoologist, follower of Cuvier and passionate opponent of Darwin, who had taught a number of them. Surprisingly enough, even they were not particularly familiar, at least initially, with Lamarck's own work. They discovered him only after evolutionary ideas had begun to take shape. One of the most prominent American neo-Lamarckians, Alpheus Packard (1839–1905), a pioneer of invertebrate embryology, published the first biography of Lamarck—in his eyes 'the founder of evolution'—in 1901. Two other leaders of the movement, Edward Cope (1840–97) and Alpheus Hyatt (1838–1902), were both palaeontologists, and they shared Packard's interest in embryology. They emphasised the active adaptation of organisms to their

Lamarck's legacy

environments; hence they reasserted the priority of function over form which had been so fundamental a part of Lamarck's transformism. Furthermore, they saw purpose in the universe as evidence of divine presence. Lamarck's idea of an inherent tendency within living things to progress and respond actively to their surroundings allowed them to formulate a compromise between evolutionism and theology by avoiding natural selection with its implications of accident and chance and stressing instead the immanent spirituality of the universe.

Lamarck offered psychologists and social theorists ways of linking the physiological, mental, and cultural aspects of evolution, as he had done for Spencer. The notion of habit Lamarck employed could provide a biological account of the processes the nascent social sciences were seeking to explain, such as the progress of civilisation or the development of the human races. Thus, although Lamarck's ideas generally appealed most to those with interests in botany and invertebrate zoology, they were also attractive, most particularly in the United States, to those attempting to put the social and human sciences on a secure intellectual and institutional base.

If there were certain distinctive ways in which Americans reacted to the evolutionary debates in general and to Lamarck in particular, the same was also true of France and Germany. The French were generally hostile to Darwin: among them he received far less recognition than elsewhere. In fact, they saw *The Origin* as just another book in a tradition already familiar to them, which had already been sufficiently debated during the famous exchanges between Cuvier and Geoffroy Saint-Hilaire in the Paris Academy of Sciences in 1830. Darwin's ideas were commonly conflated with Lamarck's, and, if given a choice, many Frenchmen preferred the latter. Indeed, they also tended to see Spencer and Haeckel as more valuable than Darwin.

Lamarck

By the end of the nineteenth century, many leading biologists in France were self-avowed Lamarckians. In 1873 a new French edition of the *Zoological Philosophy* appeared, edited by Charles Martins (1806–89). In his introduction, he praised many of the features Cuvier and Lyell had found most troubling—the synthetic spirit, the doctrine of spontaneous generation (at that time the subject of a heated scientific controversy in France) and the inheritance of acquired characters. Like all commentators he praised Lamarck's taxonomic achievements. This was the one point on which all agreed, as was clear from the appearance between 1835 and 1845 of a revised and enlarged edition of the *Natural History of Invertebrates*. Martins applauded Lamarck's environmentalism, as did many French naturalists, and his positivistic spirit in stressing the importance of observation. Martins exonerated Lamarck from the charge of materialism, a sentiment echoed by other nineteenth-century followers of Lamarck.

In Germany these themes were taken up by the vociferous supporter of Lamarck, Ernst Haeckel (1834–1919), one of the earliest to support and defend Darwin in his country. He synthesised Lamarck's and Darwin's ideas and remained faithful to the inheritance of acquired characters. He therefore opposed Weismann and the neo-Darwinist movement, and, as a Lamarckian, adhered to the reality of spontaneous generation and its significance in evolution. He developed an elaborate philosophical system based on his own peculiar brand of evolutionism, which contained elements Lamarck would have had no sympathy with whatever. Haeckel conceived his 'monism', as he called his philosophy, as a compromise between Christianity and mechanistic materialism. It led him to deny that there were absolute differences between organic and inorganic substances. Lamarck would not have approved.

Lamarck's legacy

Developments in genetics had a profound effect on biology, though the impact was far from instantaneous. When Weismann attacked the inheritance of acquired characters by claiming that all inheritance took place through the chromosomes, he denied that there was any mechanism by which acquired characters could affect the germ plasm—the special, reproductive material. His work led to impassioned debate and prompted many experimental attempts to show that acquired characters could indeed be inherited even if the precise mechanism which enabled this to happen was not yet understood. Neither Weismann's ideas nor the rediscovery of Mendel's Laws of Inheritance in 1900 laid Lamarckism to rest. For decades to come prominent biologists continued to defend Lamarck's biological philosophy and the inheritance of acquired characters. One of the last flings of Lamarckism proper came in the 1950s with the work of H. G. Cannon (1897–1963), Professor of Zoology at the University of Manchester. In *Lamarck and Modern Genetics* (1959) he attempted to rescue Lamarck's reputation; he attacked neo-Mendelian genetics for treating organisms like mechanical objects and thereby neglecting their active powers of adaptation. Like Haeckel, he had no wish to deny the value of Darwin's discovery of natural selection, which he accepted as a proven fact. It was the postulated mechanism of natural selection that he doubted, for he found random genetic mutations inherently implausible. For Cannon, the inheritance of acquired characters was an obvious part of the active responsiveness of organisms to their environment and hence a central feature of the evolutionary process.

As a final assessment of Lamarck's legacy, three comments are in order. First, during the nineteenth century evolutionary ideas transcended scientific discourse and penetrated all

Lamarck

aspects of cultural and intellectual life. On the basis of the overall pattern of nature's history an entire world-view could be erected, and, as we have seen, there were many such competing views which embodied totally different ideological commitments. It therefore mattered precisely which vision of evolution prevailed, and this accounts for the polemical and heated nature of the debates about Lamarckism and Darwinism.

The second point follows on directly. In such a situation, scientists used a number of strategies to justify their beliefs. One of the most common of these was the use of a historical figure to legitimise their ideas. Lamarck was someone who attained the position of a hero in a mythical lineage of ideas—hence the attraction of the idea of a precursor or forerunner. Significantly enough, many Lamarckians wrote histories of science in order to 'set the record straight'. It was one way of propping up allies and demolishing opponents.

Myth-making in science is also a facet of the third issue—the variety of ways in which Lamarck has been interpreted. Since subsequent generations used him for their own concerns, they inevitably took from Lamarck what served their immediate needs, needs which arose in a context radically different from Lamarck's. This accounts for the national differences noted above. By recasting his ideas in the language of the time, they inevitably changed their meaning and significance.

Yet, in the last analysis, Lamarck must have offered something valuable to the generations of naturalists who followed him and who chose him as a hero figure. Similarly, those who singled him out for attack perceived a special threat in his ideas. No one feature of his life and work can explain such phenomena. This book has attempted to set out the elements of Lamarck's biological philosophy in order to show its richness and suggestiveness and to

Lamarck's legacy

indicate where it became the focus of debate. Since interpretations of nature underlie all scientific ideas, it is Lamarck's vision of nature and its history, complex and shifting though it may be, which is worthy of our attention, understanding and sympathy.

Further Reading

Few modern editions of Lamarck's works are available in any language. The *Philosophie zoologique*, the work with which readers should begin, is in a reprint edition (Hafner Publishing, New York, 1960), as is the only English translation of it, by Hugh Elliot, *Zoological Philosophy*, first published in 1914 (Hafner Publishing, New York, 1963). Unfortunately, this version is not reliable. Also available in English is *Hydrogeology* (see Abbreviations). Few of Lamarck's books reach secondhand shops and some are exceedingly rare. There are some French pocket editions of selected pieces by Lamarck but none of these is particularly to be recommended. A useful collection of his manuscripts has appeared recently, *Inédits de Lamarck*, edited by M. Vachon, G. Rousseau and Y. Laissus (Masson, Paris, 1972).

The standard account in English of Lamarck is F. W. Burkhardt jr., *The Spirit of System. Lamarck and Evolutionary Biology* (Harvard University Press, 1977). An informative biographical account by L. J. Burlingame may be found in the *Dictionary of Scientific Biography*, vol. 7, 584–94 (Charles Scribner's Sons, New York, 1973). Also of interest is the more philosophical *Lamarck, The Myth of the Precursor* by M. B. Madaule (MIT Press, Cambridge, Mass., 1982). The biography by the noted American neo-Lamarckian A. S. Packard, *Lamarck, The Founder of Evolution. His Life and Work* (Longmans, London, 1901) is particularly interesting on Lamarck's followers. On neo-Lamarckism more generally, there is J. R. Moore, *The Post-Darwinian Controversies* (Cambridge University Press, 1979), especially ch. 6.

Further Reading

Among general studies E. S. Russell's *Form and Function* (Chicago University Press, 1982), first published in 1916, is highly to be recommended. Still useful is J. C. Greene, *The Death of Adam. Evolution and its Impact on Western Thought* (Iowa State University Press, 1959, repr. Mentor Book, New York, 1961). B. Glass, O. Temkin and W. L. Strauss jr. (eds), *Forerunners of Darwin* (Johns Hopkins University Press, Baltimore, 1959) is rather dated in its approach but is widely available and has not been superseded.

Much of value on the French scientific community of Lamarck's time will be found in D. Outram, *Power and Authority in Post-Revolutionary France: the Career of Georges Cuvier* (Manchester University Press, 1984). Michel Foucault's exhilarating book *The Order of Things. An Archaeology of the Human Sciences* (Tavistock, London, 1970) should be consulted by those with an interest in classification, especially ch. 5.

Index

Academy of Sciences, Paris, 4, 78, 101, 103, 109
Adanson, M., 13
Agassiz, L., 108
analysis, 18, 22, 23, 36, 38, 43, 57, 93, 94
Analytical System of Man's Positive Knowledge, 9, 82, 83, 98
Aristotle, 68
atmosphere, 31, 45, 53, 61, 66, 67–8, 70

Bichat, X., 3
binomial nomenclature, 12–13, 15
biology, 1, 4, 6, 25, 26, 30–2, 44–7, 59, 60, 67–8, 81, 82
Blainville, H. M. D. de, 5
Bonnet, C., 9
Botanical Dictionary, 27
botany, 1, 3, 10, 12–13, 16, 17, 18, 21, 23, 25–32, 33, 39, 57, 65, 72, 79, 109
Bruguières, J.-G., 5
Buffon, Georges, comte de, 4, 5, 9, 12–13, 15, 33, 62, 63, 73

Cabanis, P. J. G., 78
Candolle, A. P. de, 19
Cannon, H. G., 111
catastrophes, 33, 34, 63, 65, 69
chain of being, 14–15, 31
Chambers, R., 104–5
chemistry, 1, 4, 6, 31, 44, 49, 50, 57, 64, 68–9, 70, 71

classification, 1, 8, 11–24, 25, 31, 35, 38, 40, 57, 60, 65, 68, 72, 75, 79, 81, 82, 83, 89, 90, 92, 115, *and see* taxonomy
Condillac, l'Abbé, 22–3, 59, 77, 97
Condorcet, M. J. A. N., 77
Cope, E., 108
Cuvier, G., 5, 33, 35, 98, 101–3, 108, 109, 110, 115

D'Alembert, J. Le R., 10
Darwin, C., 8, 34, 101, 103, 104, 105–6, 108, 109, 110, 111, 112, 114, 115
Darwin, E., 9
degradation, 37, 73
Descartes, R., 3, 9, 30, 94
Diderot, D., 10

environment, 1, 3, 4, 6, 26, 31, 35, 39, 40, 41, 42, 46, 48, 50, 51, 53, 55, 57, 58–70, 73, 74, 75, 76, 80, 82, 86, 90, 96, 98, 102, 103, 104, 107, 109, 110, 111
evolution, 6–7, 8, 34, 40, 98, 100–11, 115
extinction, 33, 34, 35, 63, 65, 71, 72, 86, 106

faculties, 26, 37, 41, 42, 47–8, 55, 76, 79, 81
fluids, 49–50, 51, 53, 56, 58, 61, 64, 69, 76, 82, 91, 95–7

Index

fossils, 6, 33, 34, 35, 63, 65, 70, 71, 103, 105
Fourcroy, A. de, 50
French Flora, 3–4, 12, 16, 17, 20, 21, 25, 29, 38

genetics, 40, 111
Geoffroy Saint-Hilaire, E., 5, 102–3, 109
geology, 1, 6, 34, 45, 59, 63, 64, 69–70, 71, 72, 104
Giard, A., 5
God, 12, 14, 15, 21, 43, 83–8, 89, 94, 105, 109

habit, 54–6, 80, 90, 95, 97, 100, 109
Haeckel, E., 109, 110
Haller, A. von, 9, 29, 59
Haüy, R.-J., 5
Helvétius, C.-A., 77
Hippocrates, 58
history of nature, *see* transformism
Hutton, J., 34
Hyatt, A., 108
Hydrogeology, 6, 44, 64, 65, 69, 74, 114

idéologues, 22, 38, 43, 77, 79, 89, 97
inheritance of acquired characters, 2, 80, 82, 95–6, 100, 105, 107, 110, 111
invertebrates, 6, 7, 27, 34, 36, 37, 39, 42, 43, 70, 72, 82, 93
irritability, 22, 28–30, 39, 43, 49, 51, 59, 76

Jardin du Roi (Muséum), 4, 6, 7, 13, 33, 35, 36, 71, 101
Jussieu, A. L. de, 13, 19

Lamarckism, 100, 107–10, 115
Latreille, P.-A., 5, 7
Lavoisier, A.-L., 50, 68
laws of nature, 9, 15, 24, 33, 43, 46, 47, 57, 61, 66, 71, 73, 74, 81, 83, 84, 85, 86, 88, 93, 95, 107
life, *see* biology *and* power of life
Linnaeus, C., 9, 12–13, 14, 15, 17, 20, 25
Locke, J., 9, 22–3, 59
Lyell, Sir Charles, 34, 103–4, 106, 110

man, 1, 7, 21, 22, 24, 41, 58, 59, 62, 64, 72, 77, 78, 79, 81, 82, 84, 86, 87, 89–99, 103, 107
Martins, C., 110
Mendel, G., 111
Meteorological Annuals, 4, 65, 67
meteorology, 1, 3, 4, 45, 59, 60, 65–8, 70
mineralogy, 26, 62, 70
Muséum, *see* Jardin du Roi

Napoleon, 4, 67, 78
Natural History of Invertebrates, 8, 34, 39, 41–2, 82, 83, 110
nervous system, 8, 38, 39, 41, 42, 50, 72, 76, 77, 79, 82, 84, 91–2, 95, 97, 102
Newton, Sir Isaac, 13, 49

Olivier, G.-A., 5
organisation, 20, 30, 37, 38, 40, 46, 51, 54, 73, 74, 77, 78, 79, 92, 93, 97, 102

Packard, A., 108, 114

117

Index

Perrier, E., 5
physics, 4, 31, 44, 49, 57
power of life, 42, 43, 48, 64, 73, 75, 82
psychology, 1, 24, 31, 81, 84, 93, 109

Ray, J., 11
Researches into the Causes of the Principal Physical Facts, 89
Researches on the Organisation of Living Bodies, 8, 54, 73, 80, 89
Rousseau, J.-J., 9, 92, 94

scientific method, 11–24, 47, 65, 66, 69, 83, 88, 89, 94, 97, 99
sensation, 21, 22, 39, 72, 76, 84, 96–7
sensibility, 21, 28–9, 59, 102
series, 14, 20, 37, 39, 42, 56, 64, 73, 75, 83
Spencer, H., 106–7, 109
spontaneous generation, 46, 47, 52, 62, 71, 106, 110
structure and function, 29–30, 82, 92, 102
Synopsis of Plants, 19

System of Invertebrate Animals, 8, 36, 38, 73

taxonomy, 1, 7, 9, 26, 27, 33, 35, 40, 41, 60, 73, 79, 84, 93, 98, 99, 101, 105, 110, *and see* classification
Tournefort, J. de, 13
transformism, 8, 10, 15, 22, 28, 31, 34, 35, 47, 51, 54, 56–7, 64, 65, 70, 71–82, 83, 84, 85, 90, 96, 98, 103

uniformitarianism, 34, 104

vertebrates, 36, 37, 39, 42, 43, 76, 82
vital principle, 21, 42, 44, 45, 84, 87, 88, 89

Wallace, A. R., 108
Weismann, A., 100, 108, 110, 111

Zoological Philosophy, 8, 31, 37, 38, 57, 67, 71–82, 83, 97, 110, 114
zoology, 1, 17, 26, 30, 31, 33–43, 57, 75, 78, 79, 82, 93, 97, 109